Geography
Our Physical and Human Resources

Brian Knapp
David Worrall

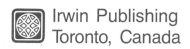
Irwin Publishing
Toronto, Canada

This Canadian edition is derived from *Earth and Man* by B.J. Knapp, Unwin Hyman Limited, London, UK.

Copyright © 1989 Irwin Publishing

Canadian Cataloguing in Publication Data

Knapp, B.J. (Brian John)
  Geography : our physical and human resources

Includes index.
ISBN 0-7725-1714-2

1. Physical geography.　I. Worrall, David Spencer.
II. Title.

GB55.K63 1989　　　910'.02　　　C89-093074-0

Cover photo: Copyright Advanced Satellite Productions Inc. 1988/Satellite image of southwest BC is supplied courtesy of Advanced Satellite Productions Inc., Vancouver, BC.

Cover design: Jack Steiner
Illustrations: Catherine Farley
Typesetting: Jay Tee Graphics Ltd.

No part of this publication may be reproduced or transmitted in any form or by any means, electronic or mechanical, including photocopy, recording, or any information storage and retrieval system now known or to be invented, without permission in writing from the publisher, except by a reviewer who wishes to quote brief passages in connection with a review written for inclusion in a magazine, newspaper, or broadcast.

1 2 3 4 5 FP 93 92 91 90 89

Printed and bound in Canada by DW Friesen Printers

Every effort has been made to trace the original sources of material and photographs contained in this book. Where the attempt has been unsuccessful, the publishers would be pleased to hear from copyright holders to rectify any omission.

Preface

Geography is the study of the earth as our home — the study of landscapes, environments, and people. Whenever we move about our neighbourhoods, towns, cities, or travel even farther, the opportunities for appreciating geography are all around us. Geography is a field study. In fact, the geographer's approach is in many ways like that of a detective — a modern-day Sherlock Holmes. Like the investigator, the geographer is rarely able to witness the events in progress. Rather, the geographer goes to the "scene of the crime" and begins to assemble the evidence in order to reconstruct the events. The plan of a river, some rock fragments, the shape of the slopes, the location of a town, the colour of the soil, and the nature of the vegetation are just a few clues. Evidence is assembled, arranged, and many possible explanations are developed. Ideas and generalizations are tested. Some are rejected and others are held for further testing as new evidence continues to accumulate. The results of generations of detective work make up the contents of this book.

One of the main purposes of this book is to help you to better understand and appreciate your surroundings — the form and functions of the earth, its topography, processes, and resources. Another purpose is to help you prepare for the provincial examination in Geography 12. The authors hope that you will develop a keen interest in the causes and effects of the changes that surround you; changes that make this earth such a fascinating home. Your interest will help you learn effectively so that the examination will be, as Holmes said, "Elementary, my dear Watson!"

D.S. Worrall *B.J. Knapp*
Kelowna, BC

Acknowledgements

The rewriting of *Earth and Man* has been a labour of love made possible by the agreement of Dr. Brian Knapp, the work of a dedicated team of professionals, and the modern technological innovations of the FAX and Priority Post.

Through the autumn of 1988 when two chapters were being rewritten each week, and every deadline met, no one gave me more support and encouragement than my wife, Carmel.

Maryjean Lancefield as Project Editor maintained an almost impossible schedule by providing energetic technical and artistic support and maintaining excellent communications and encouragement throughout. Judith Lynn of Irwin tirelessly and with unfailing good humour typed and photocopied the manuscript and the voluminous correspondence.

The excellent, thorough review work of Mary Norris (Victoria) and the outstanding technical and pedagogical analyses of John Chalk (Principal, Templeton Secondary School, Vancouver) contributed immensely to the text. Thelma Brooks, John Chalk, and Harvie Walker, fellow members of the BC Geography Curriculum Revision Committee, provided critical comment upon the text, glossary, and the extent to which this volume addressed the intended learning outcomes of the new curriculum.

Catherine Farley, the illustrator, converted my rough sketches into works of art which should make geography more memorable and entertaining to our students.

Contents

Preface	iii
Acknowledgements	iii
Starting points	1

Part 1   **THE ORIGIN OF LAND AND SEA**   5

1	The origin of the earth's rocks	6
2	Rocks made from sediments	23
3	Earthquakes, faults, and folds	33
4	Continental drift and mountain building	40

Part 2   **ATMOSPHERE AND OCEAN**   49

5	The weather machine	50
6	The world's weather	71
7	The waters of the oceans	81
8	The water cycle	89

Part 3   **SOILS, VEGETATION, AND PEOPLE**   107

9	Weathering	108
10	Soil formation	114
11	Clothing the surface	121
12	Cultural ecosystems	138

Part 4   **SHAPING THE LAND**   157

13	Mass movement processes	158
14	Rivers and valleys in humid regions	170
15	Running water and limestones	191
16	Deserts	195
17	Glacial landforms	203
18	Coastal landforms	219

Appendix 1	Latitude and longitude	234
Appendix 2	Time	235
Appendix 3	Instruments to measure things	235
Glossary		238
Sources and credits		246
Index		247
Key to multiple-choice questions		250

Starting points

The study of geography

Dying for a drink

Figure A Sudanese herders collecting water from a shallow well dug in a river bed

Did you know that each year more people die from drinking water than alcohol? As you read this page hundreds of thousands of people will have to trudge several kilometres to their nearest water supply (Fig. A); it will be a hole in the ground. They will fill old gas cans and leather buckets, trudge back home and then drink the dirty water they have collected. The water will make many ill and some will die. But why should some places have so little water and other resources, and how does this affect the way their people live?

Did you know that, one morning a few years ago, a whole school in a Welsh valley was buried under a landslide and 145 children were killed? Why should such accidents happen? What are the forces that cause landslides? Can they be predicted or prevented?

Why is it unsafe to build near some rivers? Why does a road have to go down a slope and what created the shape of the distant hills? How does the shape of land beneath your feet affect your life?

If you have ever asked questions like these about people and their environments you will have been asking questions about geography. This book tries to answer some of the common questions and to help to put the pieces of the environment puzzle into some sort of order. But geography is a wide subject, and here the emphasis is on the natural environment — air, sea, plants, soil, and rock. Using our understanding of the natural environment we then try to show some of the ways in which the environment has affected the way people have lived in the past and live today. Today, more than ever, we need to understand the natural processes at work on the earth's surface so that we can help make good use of our environment, not destroy it or be destroyed by it.

Often, starting points for interest in the environment are the stories of human tragedy that make the headlines. These are events where people and environments are in dramatic conflict. However, in general, more subtle processes are at work, going unnoticed because they operate slowly. On the following pages we show how a real landscape can be a mixture of such subtle processes.

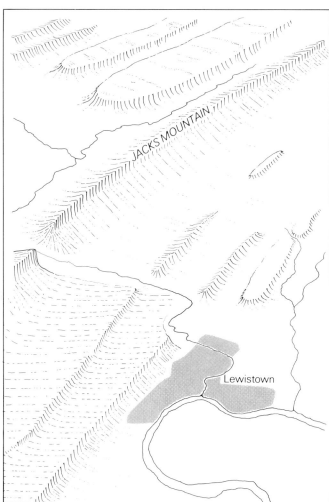

Figure B An area of ridge and valley scenery, Pennsylvanian Appalachians (USA)

Topic   Air photos, maps, and sections

There are many photographs in this book, some taken from the ground and others from the air. Some are taken from satellites and yet others by astronauts in space. These photographs give us different views of the earth's surface and they are very valuable in helping to build an accurate picture of the landscape. However, photographs can be deceptive, and the geographer must make extensive use of maps, cross sections, and detailed field work to complete the picture.

Figure C is a map of the area shown in Figure B. Here the height of the landscape is shown by contour lines. The spacing of the contours tells us about the steepness of the slopes. Because the ridges, for example, have contours spaced more closely on one side than the other, we know the steeper slopes face southeast. The cross section (Fig. D) also helps to show the main features of the landscape in a view from the side. In the cross section, the steeper southeast slopes show up very clearly and we get a better impression of the relative steepness of the slopes and the shape of the valleys. We can also see how landscape and rock types are closely related.

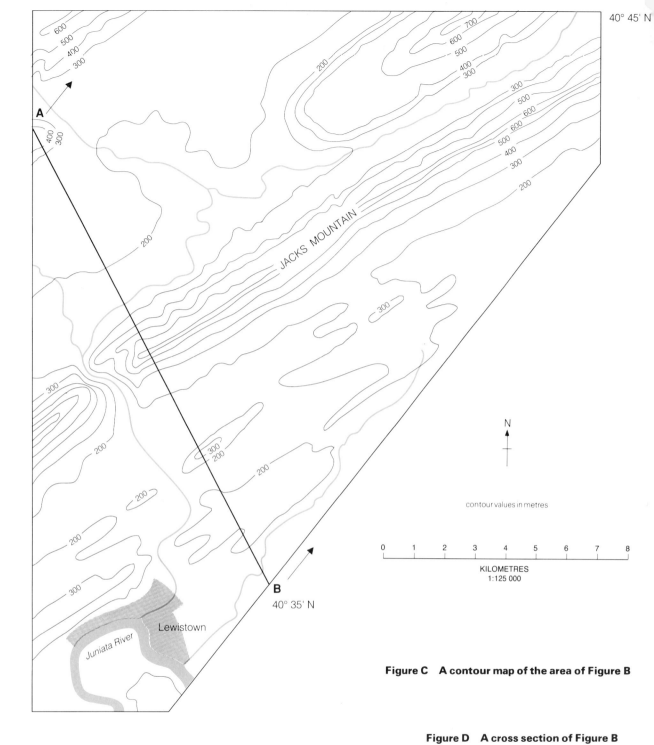

Figure C A contour map of the area of Figure B

Figure D A cross section of Figure B

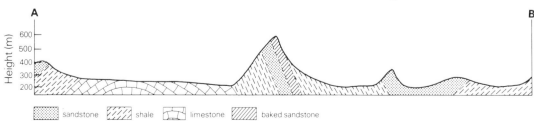

The people-environment relationship

Earth and people

Look at Figure B. The photograph was taken from a plane and it shows some common relationships between earth and people. A long, dark, forested ridge cuts across the landscape like a knife, separating two areas of snow-covered lowland. A small river flows through a deep cleft in the ridge and then twists in its path to flow through two more gaps in parallel ridges. Eventually it joins the large river in the foreground.

This is the natural landscape. It owes its shape to the rocks below the surface and the delicate etching of water and ice which have emphasized differences in the resistance of the rocks. But what causes some rocks to be more resistant than others? How did the rocks form, why were they tilted, and exactly how were they attacked by water and ice?

People have settled in the natural landscape. They must have been influenced by the gaps in the ridges when they first made roads, laid out railways, and built towns. Clearly the gaps serve to focus railways, roads, and people, as well as the river system. People have not settled on the upland ridges which remain forested and uninhabited. Instead cultivation of the more fertile lowland areas has produced a pattern of fields, where trees once grew. But how dependent is the agricultural system on the climate of the area and the soils in which the crops grow? What restrictions do they impose?

The large river has also played its part. It has influenced the site of the town because it provided a means of carrying goods. As the town grew away from the river its streets were laid out in straight, unnatural lines, but the direction of development has still been strikingly influenced by the ridge and the river, forcing it to squeeze in between them.

The area in the photo is typical of the way in which people have taken advantage of the landscape. As technology advances, so people have a greater impact on the environment and are less at its mercy. They can build tunnels through mountains where beforehand a natural pass had to be found between them. People can harness rivers to supply energy, prevent floods, and irrigate vast areas of desert. Nevertheless nature cannot be ignored, nor can each part of the environment be looked at in isolation and without a plan. We must study the earth in a systematic way so that, at the end, we can see how all of the parts are related. In the past too many important environmental decisions have been taken with little understanding of their wider effects. Some will be described in later chapters. Examples are wide ranging and they include a dam, built for hydro-electric power, that caused an earthquake, a landslide, and a flood; and aid from developed countries to help irrigate parts of the Indian subcontinent that killed the very plants it was intended to save. Geography can make an important contribution to the welfare of people by helping us to see the complex way in which the environment works. So in our study of the earth's patterns and the processes that cause them we have very good reasons for starting geography with our feet on the ground–and with the rocks beneath our feet.

Part one The origin of land and sea

The materials of the earth's surface are continually being eroded, renewed, and rearranged.

Figure 1.0 Mt. St. Helens volcano erupting

1 The origin of the earth's rocks

Evidence of earth history

Ancient rocks

In 1966 a young New Zealand scientist working in Greenland found some unusual light and dark blotchy rock. When he sent samples of the rock for analysis they were found to be the most ancient rocks ever discovered — 3800 million years old!

There are also small areas of very old rocks elsewhere. Rocks in the American state of Minnesota, for example, have recently been found to be of similar age to those in Greenland, and in South Africa there are rocks that formed 3500 million years ago. The South African rocks are also very interesting because they are made from muddy debris eroded by rivers and deposited on the floor of the ocean. This tells us that we have had land and oceans, clouds, rain, and rivers for at least 3500 million years. Indeed, the earth's surface must have looked much as it does today, except that there was no vegetation or animal life of any kind on the land.

If the earth's surface has looked familiar for so long, how old is our planet? Most scientists agree that the earth developed out of a cloud of "cosmic dust" about 4500 million years ago — about the same time as our neighbouring planets and the sun. This has been confirmed by dating the lunar rocks (brought back by astronauts) and the meteorites that occasionally come crashing through our atmosphere.

Nevertheless, although dating the birth of the planet is becoming possible, we are still far from understanding how it formed. Scientists are still arguing whether the space dust suddenly crashed together or whether it came together more slowly, picking up space debris such as meteorites as it spun through our part of the galaxy. Whichever process was responsible, it is clear from the outpourings of volcanoes that deep inside the earth the rocks became very hot. Some would argue that an initial crash of dust could have generated enough heat to melt the earth's interior, but, as that was over 4500 million years ago, it can hardly be important today. However, we may get a clue to present processes by looking at the processes operating in a nuclear power station. Nuclear power stations use the heat given out by radioactive substances such as uranium when they change from one form to another. There are many such substances contained within the earth's rocks whose chemical changes could continuously fuel the earth's internal cauldron. But their significance goes much farther than just providing heat because, as we shall see, rocks from molten materials have provided us with the foundation of the earth's surface rocks.

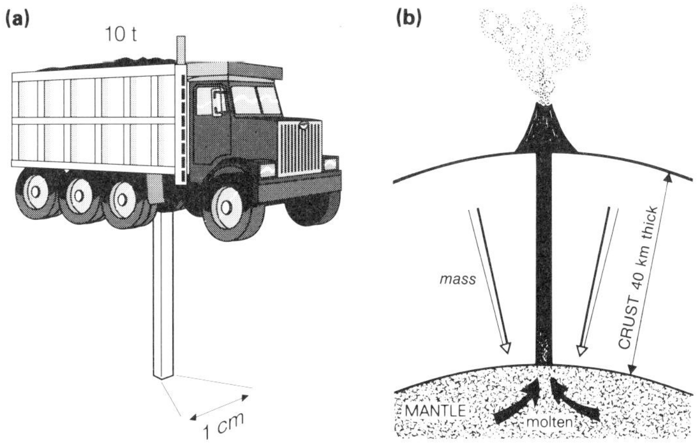

Figure 1.1 Volcanic eruption (a) At 40 km below the earth's surface the pressure is nearly 10 t/cm² as a result of the mass of the overlying rock. (b) The immense pressure deep in the earth soon forces out molten rock if a line of crustal weakness develops.

Rock under pressure

Solid → liquid → gas

With a cool surface and very hot interior, there must be enormous changes in rock temperatures with depth. Indeed, near the surface, temperatures often increase by 30°C every kilometre. This change is even noticeable in deep mines where miners often have to work in "tropical" temperatures.

Although this high rate of temperature increase is not maintained deep into the earth, it would still mean the rock should become hot enough to melt at just a few tens of kilometres below the surface. We cannot prove this directly, but the evidence indicates that molten conditions only occur very close to the centre of the earth, thousands of kilometres below the surface. Above this the great weight of overlying rock provides enough pressure to keep most of the earth solid. But, if the surface rocks crack and the pressure is eased, there is an opportunity for the greatly heated rock to become liquid (Fig. 1.1). Any subsurface rock that becomes liquid in this way is called **magma**. As magma reaches the surfaces it pours out as spectacular flows of **lava**. At the same time, some parts of the liquid boil off as gas and help form the huge plumes of cloud that tower over a volcanic eruption (Fig. 1.0).

Convection within the earth

Fire burn and cauldron bubble

Most probably at its beginning the earth did not have a separate solid surface of low density rocks and a molten core of much heavier materials. However, with time, heat built up and caused the centre to melt, but the outside became solid and much colder, constantly losing heat to space. As a result a temperature difference developed between the surface and the core sufficient to cause a slow, circulating movement **(convection)** of the molten rock in the earth. In some ways the process may have been similar to heating a saucepan full of dark, thick molasses (Fig. 1.2). Although extremely slow, over many hundreds of millions of years these convection currents may have heaved the whole fabric of the earth's interior round and round, helping to separate out heavy and light materials: the light materials would be inclined to float to the surface as a sort of scum, and the heavy materials would tend to sink, thereby dividing the earth's interior into a series of shells, one inside the other (Fig. 1.3).

Figure 1.2 Earth convection
When a saucepan of molasses is heated, convection currents develop and the molasses is slowly churned around in a manner similar to that of rocks within the earth.

Figure 1.3 Structure of the earth
(a) The earth is composed of a series of rocky shells: the outer shell is the rigid *crust* a few tens of kilometres thick; the *outer core* (about 3300 km in radius) is molten; the *mantle* is solid, but moves very slowly under extreme force. (b) The earth's crust is therefore relatively thinner than the shell of an egg.

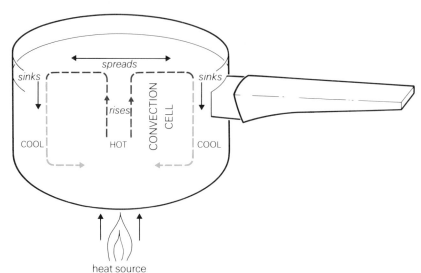

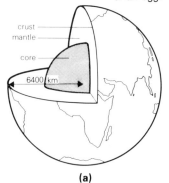

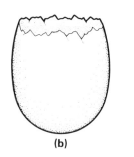

Volcanoes on Iceland

At present this "scum" layer is up to 100 km thick. We call it the **crust** of the earth and it is the solid rock on which we live. It is thickest under mountains and thinnest under oceans. However, compared with the radius of the earth — nearly 6400 km — the crust is thinner than the shell of an egg (Fig. 1.3), and just like an egg shell the crust is strong but brittle and can be cracked. As we shall see, it is along the cracks that some of the interior rock escapes and volcanoes are born.

Once the crust had formed, it may have acted as a skin of insulating material, because solid rock is a very poor conductor of heat. If this were not the case the earth's surface would still be burning hot, or alternatively the heat of the earth might have been lost entirely. In either case, life on earth would be impossible today. As it was, with only a quarter of earth history gone, the surface seems to have cooled enough for oceans, clouds, and rain to form and even for the first primitive life forms to develop.

The crust has remained floating on the **mantle** of denser slowly moving rock below like styrofoam floats on a swimming pool. Nevertheless, the vast outpourings of lava from volcanoes that occur quite frequently even today tell us that the "cauldron" remains beneath our feet, its molten rocks awaiting any opportunity to escape.

Island "fireworks"

The oceans of the world are dotted with thousands of islands, many of which have one very surprising feature in common: they are dominated by volcanoes. For example, Japan's major city, Tokyo, is overshadowed by Mt. Fuji, and Lanzarote (part of the Canary Islands, 150-300 km off the coast of North Africa) has a mountain whose slopes are still covered with cindery volcanic boulders (Fig. 1.4). Similarly all of the islands of the Caribbean Sea owe their existence to volcanic outpourings.

A volcanic eruption is one of the most spectacular and frightening sights on earth. One recently active volcano was on the Icelandic island of Heimaey in 1973. A reporter who saw the first stages of the eruption gave this vivid account:

❝A new vent has now appeared on the north-west side of the long dormant Helgafell volcano. Great clouds of debris, hot lava and black ash have been set free and now cover much of the island of Heimaey. The eruption created a rift nearly 3 km long across the eastern edge of the island, throwing up a torrent of boiling lava. Tonight the eruption is growing in intensity and the rift is still developing, northwards now, beneath the sea.

The refugees, many of whom were ferried to the mainland by air, say they have given up all hope; they fear they may never be able to return to their homes. Tonight Helgafell is angrier than ever and geologists are aware that it could be weeks, even months, before they can judge if the island can ever be reinhabited...

(and three months' later)... For the people of this island this eruption was totally unexpected. Though only a tenth of the town was destroyed by lava, the remaining buildings are now collapsing under the weight of the volcanic dust that constantly falls. The eruption has continued for more than three months and it could be more than three years before it ends, no one knows...❞ (Based on British Broadcasting Corporation News reports)

In the Atlantic Ocean most of the active volcanoes (such as those on Iceland) form a north-south line through the centre of the ocean (Fig. 1.5). But although the eruptions in the Atlantic can be violent, in general the materials that flow from these volcanoes do so relatively quietly. The lavas are mostly very dark in colour and they flow very freely out of **fissures**. Lavas of this kind are called **basalts**. Only rarely do they build up large steep-sided cones.

Free-flowing material can often force its way in between layers of rock as well as pushing to the surface. When this happens sheets of magma are injected. Although the result is not immediately visible, the long-term results on the landscape may be important. These sheets often make much harder rock than the layers they pushed between. The Fair Head Sill in Northern Ireland shows the edge of just such a sheet. Sheets of cooled magma that push between rock layers are called **sills**; if they cut across rock layers they are called **dikes**.

Figure 1.4 Cinder cones When solid rock is ejected from a volcano it builds up into cinder cones. These lumps of material are fist sized.

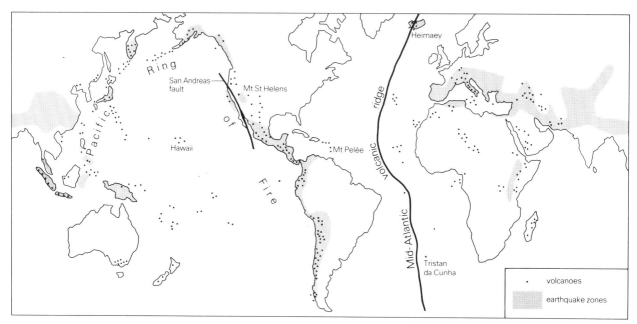

Figure 1.5 The distribution of volcanoes and earthquakes around the world

The magma that fed the Helgafell volcano has now cooled in its fissure, producing a new dike. Indeed, sills and dikes are most commonly made of basaltic material because it flows so readily. For example, Iceland is made up of many thousands of sills, dikes and lava flows, all cutting through one another and piled up in a giant mound some 6000 m high. What we see of Iceland is only the tiny exposed tip of this huge volcanic mountain.

Topic What are rocks and minerals?

The whole outer shell of the earth is made from solid rock. Sometimes we see it exposed in cliffs or mountains, although it usually lies hidden below soil and vegetation. The first rocks to form on the earth's surface were cooling magmas. Rocks which have solidified from magma are called **igneous** rocks (the word igneous comes from the Latin word for fire). Granite is an example of an igneous rock. Here the chemicals in the magma formed into substances of fixed chemical content called **minerals**. Quartz, feldspar, and mica are examples of minerals found in granite. These minerals then solidified in the form of interlocking crystals (Fig. 1.6c).

Eventually some of the igneous rocks were eroded by water, wind, and ice, and the small particles of eroded rock (called **sediment**) were carried away and deposited in the sea. These particles were eventually buried by more sediment and were slowly compacted into new rocks. These are **sedimentary** rocks. For example, sand particles on a beach can be compacted to form sandstone rock (Fig. 1.6b). Similarly, clay particles eroded from land by rivers settle out as mud on the sea floor and compacted to form shale rock.

Sometimes the burial is so deep that the rocks are crushed and partly melted until they take on a new form. These (changed) rocks are called **metamorphic** rocks (Fig. 1.6c). (See also p. 44)

There are about 2000 different minerals on the earth and in various combinations they make up a wide variety of rocks.

We use such names as granite and sandstone to describe groups of rocks that are similar, but each rock within a group has its own special features, thereby helping to give variety to our landscape.

Figure 1.6 Rock types

Through a microscope we can see (a) a piece of granite made from crystals that interlock, (b) a piece of sandstone made from rounded sand grains, (c) a piece of metamorphic gneiss with all of its crystals melted into each other.

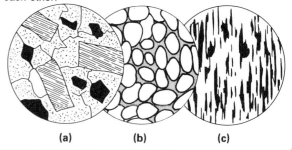

(a) (b) (c)

Volcanoes and mountains

The ring of fire

In contrast to Iceland, the many volcanoes that form a "ring of fire" around the Pacific Ocean (Fig. 1.5) are made from types of lava that flow much less readily. One main type is called **andesite** after the Andes Mountains where so many of the volcanoes are found. These volcanoes produce steep-sided cones which are mostly composed of ash rather than lava. Although they are usually smaller than those that form in the oceans (Fig. 1.7), their violence, coupled with their closeness to many centres of population, makes them some of the most dangerous natural features on earth.

Because this magma is more sticky than the basaltic type, dikes and sills are less common, but under each volcano is a vast chamber of magma deep in the crust. The pressure builds up in this chamber before a new eruption. But when the volcano finally ceases to erupt and there is little pressure in the chamber, the magma solidifies into a rock called **granite**. British Columbia's Coast Range marks the top of a complex of several great intrusions of magma. These masses are called **batholiths** (Fig. 1.17).

There are many other examples: in Africa (e.g. much of the Jos Plateau, Nigeria; in North America (e.g. the Sierra Nevada of California); and in Australia (e.g. the Snowy Mountains and the Great Dividing Range).

The Mt. St. Helens eruption

Terror, 20 km high

Quite recently we have been able to observe a violent volcanic event similar in size to the famous eruption of Vesuvius in Italy in AD 79.

In May 1980, after two months of rumbling, the beautiful ice-clad peak of

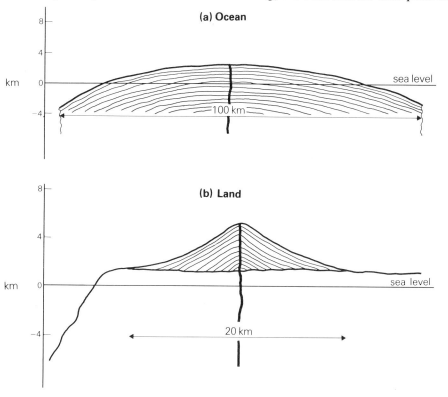

Figure 1.7 The shapes of volcanoes
Volcanoes vary in shape and size depending on their location and how easily the lava flows.
(a) Mobile basaltic lava,
(b) sticky andesitic lava.

Figure 1.8 The eruption of Mt. St. Helens volcano

Mount St. Helens in Washington State, USA, suddenly erupted (Fig. 1.8). Early on a Sunday morning an earthquake shook the area and this was followed immediately by an explosion inside the volcano equivalent to 10 million t of TNT. A nuclear bomb of this capacity would destroy Greater Vancouver.

Over 6 km³ of rock were blasted away, lopping 400 m from the top of the mountain (a piece of rock about the size of Burnaby Mountain, Vancouver). 400 km² of forest (about twice the area covered by Lake Okanagan) were destroyed. A helicopter pilot sent out to look for survivors reported "all the trees were lying parallel on the ground like big spokes emanating from the centre of the mountain... the whole area was burned over and I just couldn't believe it. It looked like the surface of the moon" (Fig. 1.9).

Figure 1.9 Mt. St. Helens *Left.* The eruption blew off a whole section of the mountain and, *right,* the blast also destroyed a vast area of forest and stripped the branches from trees. The tree trunks shown here are over 30 m long.

Next a column of sulphurous ash belched 20 km into the sky, turning day into night as it blocked out the sun. Mudflows were triggered by millions of tonnes of ice and snow which thawed instantly in the tremendous heat. The mudflows, moving at 60 km/h tore out bridges, and buried cabins. Walls of water 3 m high hurtled down the nearby rivers carrying homes, cars, and hundreds of logs in their path (Fig. 1.11).

Soon the whole air became foul for hundreds of kilometres around, carrying the acrid stench like a burned-out building. On the same scale the smell and ash cloud of a volcano erupting on Vancouver Island would reach Kamloops and Kelowna. There was five hundred times more dust and grit in the air than normal, so that face masks became a necessity. One bank in a nearby town even put up a sign: "For security reasons please remove face masks before entering the bank!" The Columbia River was choked with mud and ash, and 33 ocean-going ships were trapped in the harbour at Portland. Indeed, the prevailing westerly winds slowly sent a huge cloud of ash right across the United States and out over the Atlantic Ocean (Fig. 1.10).

Ash is an important part of a violent volcanic eruption and it is often much greater in volume than any lava. In the 1980 eruption, Mount St. Helens produced hardly any lava. Notice also that the eruption triggered off other natural disasters such as mudflows, which reminds us that the effects of a volcanic eruption are not just confined to the ejection of lava and ash.

Figure 1.10 Mt. St. Helens The extent of the ash cloud one week after the eruption.

Figure 1.11 Mt. St. Helens Mudflows and melting ice caused severe flooding in the nearby valleys.

Exercise 1.1 The nature of lava flows

We have maps of Vesuvius (near Naples in southern Italy) showing the lava flows for nearly 300 years and this provides an excellent opportunity to study the effects of the eruptions (Fig. 1.12). Notice that the top of the cone is "lopsided" and for many hundreds of years lava has flowed out mainly towards the south.

1. Choose one of the long lava flows with a date on and make a copy of it. Note that this lava is of the sticky type and it does not flow very far. What shape is a typical lava flow?
2. Write down the date of every lava flow in order, putting the earliest first. Put a cross next to each date that had a particularly large flow. Is there any regularity or pattern in the dates of eruptions?
3. Do the larger lava flows correspond with the greatest intervals between eruptions?
4. There has not been a major lava flow for 50 years. Do you think one is likely soon and will it be large or small? Note that before the famous AD 79 eruption the volcano had been inactive (**dormant**) for so long that the Romans thought it was extinct. Also Pompeii was buried in falling ash while nearby Herculanium was destroyed by a mudflow caused by a mixture of torrential rain and ash. Neither town was engulfed by lava. After the eruption of AD 79 the volcano was quiet again until AD 172.

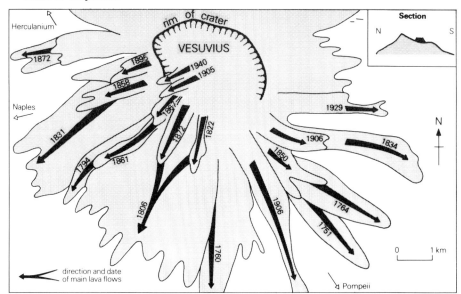

Figure 1.12 The eruptions of Vesuvius

Figure 1.13 The Helgafell volcano
The lava and ash deposits that partially covered Heimaey, Iceland.

Exercise 1.2 The deposits of an active volcano

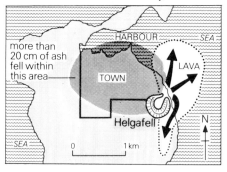

1. Turn back to page 9 and read the description of the eruption of Helgafell, a volcano on an Icelandic island. Also study the map (Fig. 1.13). What were the two main products of this eruption? Which was the most extensive?
2. The map shows the areas affected by the eruption of the Helgafell volcano. The eruption was partly from a long fissure. How did this affect the distribution of lava?
3. At the time of the eruption the wind was blowing from the east. How does the distribution of ash deposits tell us this?
4. What would have been the least hazardous direction for the wind to have been blowing from at the time? Why?
5. The description tells us that a volcanic eruption produces more than one type of material. Describe the differences between the distributions of lava and ash.

Topic Dikes, sills, and batholiths

Dikes are sheets of igneous rock which cut across other rocks. Very often they are harder than the rocks they cut, and erosion leaves them standing up as ridges (Fig. 1.14). Dikes can be seen clearly around Ship Rock, New Mexico, an old volcanic plug (Fig. 1.15). Sills are made from the same materials as dikes and are sheets of igneous rock. However, sills were injected between, rather than across, other rocks. As with dikes, sills can be more resistant to erosion than are other rocks. Hadrian's Wall, an ancient Roman wall in northern Britain, was built on the Great Whin Sill because it was prominent as a long, natural ridge in the landscape. High Force, a famous British waterfall, has been formed by the same sill (Fig. 1.16). Batholiths are large magma chambers that once fed volcanoes but are now filled with igneous rock. They are usually made from granite and they provide the largest intrusive landscape features. Much of the Coast Range of British Columbia and the Sierra Nevada in California are granite batholiths (Fig. 1.17).

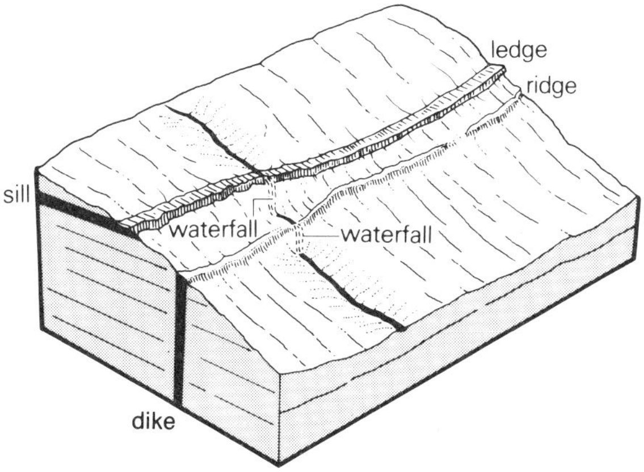

Figure 1.16 Sills
The River Tees in Durham (England) flows over the Great Whin Sill, forming a spectacular waterfall.

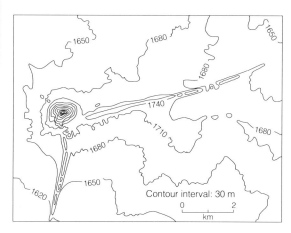

Figure 1.14 Intrusions
Many small dikes and sills form ridges and benches in the landscape.

Figure 1.15 Ship Rock, New Mexico
Navajo legend tells us that Ship Rock is all that remains of a great ship that brought the Navajo from northern lands. The residual volcanic plug and radiating dikes are spectacular local features.

Figure 1.17 A batholith
Yosemite Valley is a magnificent glacial trough which has been eroded into the massive granite batholith which forms this section of the Sierra Nevada, California.

SUMMARY: IGNEOUS ROCKS

Igneous rocks are those that have solidified from **magma** as it cooled on or near the earth's surface. Beneath the earth's rigid **crust** there is a mixture of solid and partly molten rock. When a line of weakness forms in the rigid outer crust, the magma nearer the surface could possibly erupt as a **volcano** (Fig. 1.18). If it remains in liquid form the magma is called **lava,** but often the magma boils off as gas or is blown out of the volcano in granular solids called **ash** or in larger lumps called **cinders.** Much of the time magma cannot push its way to the surface either because it is too sticky or because there is very little pressure from below. In these cases the magma will solidify within the crustal rocks.

Magma that reaches the surface cools quickly in air and forms only tiny crystals, whereas buried magma cools much more slowly and forms larger crystals. Lava is called an **extrusive** rock because it is extruded (forced out) onto the earth's surface. Basalts are lavas that are very runny; andesites and rhyolites are sticky lavas that help make steep volcanic cones (Fig. 1.0). Some of the magma is usually trapped underground and may then be intruded (forced between) sheets of cold rock. These sheets of **intrusive** rocks are called dikes, if they cut across the cold rocks; or **sills**, if they form parallel to the sheets of cold rock. Eventually even the magma chamber that supplied the materials for the volcanoes, dikes and sills may cool down. These magma chambers can be very large — sometimes hundreds of cubic kilometres in volume. When they cool they form granite rocks. A solidified magma chamber is called a **batholith.**

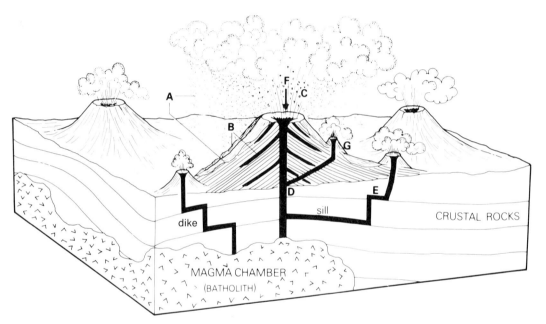

Figure 1.18 Characteristic forms of igneous activity
Write down the names for the features labelled A-G.

The Cascades Mountains

Fiery Cascades

Mount St. Helens is just one of a whole series of volcanic peaks that form the Cascade Range of the Western USA (Fig. 1.19). It is the most active, but nearby there are volcanoes that must have been even more stupendous in their day. One of the largest of these "inactive" volcanoes contains the famous Crater Lake (Fig. 1.20). About 6000 years ago this volcano was over 4000 m high and it erupted in a long series of fiery blasts. It threw 60 km^3 of incandescent ash and pumice into the air — ten times the volume blown from Mt. St. Helens! After it had blown so much **pyroclastic material** into the air and sent avalanches of ash, pumice, and lava down its slopes, the unsupported cone collapsed, so that at the end of the eruption the high ice-clad peak had vanished and in its place was a vast pit about 8 km in diameter and 1.5 km deep. Later a very small cone was built up by a small eruption, while rain and snow water gradually filled the crater until the cone was only a small island in a lake. This very violent explosion probably removed the magma from the magma chamber faster than it could seep in from the mantle below. With the chamber partly emptied, the support for the volcano was largely removed, allowing the mass of the cone to cause the spectacular collapse (Fig. 1.21). Collapsed volcanoes are known as **calderas**.

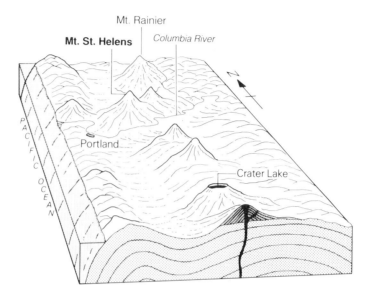

Figure 1.19 Cascade Range, Washington State, USA
These rugged mountains, part of the volcanic formations near Mt. Rainier, rise above the deeply dissected plateau of the northwestern USA.

Figure 1.20 Crater Lake
This lake now occupies the centre of a volcano that blew its top only 6000 years ago. However, the most recent volcanic activity was about 1000 years ago. The small secondary ash cone that forms Wizard Island in the lake tells us that there might be more violence to come.

Exercise 1.3 Caldera formation

Figure 1.21 summarizes the development of Crater Lake. The diagrams and captions are not in correct order. Match the captions with the diagrams, and write them in the correct order.

(v) The volcano begins its final series of eruptions with an upwelling of magma and a light rain of incandescent pumice.

(w) Activity is greatly reduced and a lake forms in the caldera. A brief period of rebuilding results in the formation of a small cinder cone. The cliffs slowly weather enlarging the caldera.

(x) The cone of the volcano collapses (implodes) into the magma chamber creating a huge crater (caldera) within the rugged remains of the mountain.

(y) The violently active volcano blasts almost 60 km^3 of pyroclastic material, mostly ash and pumice, high into the air. Most debris avalanches down the side of the mountain, but ash falls over an area of 900 000 km^2. The magma chamber empties rapidly.

(z) Very vigorous gaseous and ash eruptions empty the magma chamber faster than it can be recharged from below. Support for the cone is reduced and cracks develop in the upper parts of the mountain.

Figure 1.21 Calderas
(a-e) The stages leading to the formation of the famous caldera now occupied by Crater Lake, Oregon. The great eruptions which led to the implosion of the cone spread volcanic ash over most of the western US and southwestern Canada.

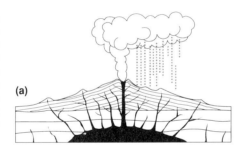

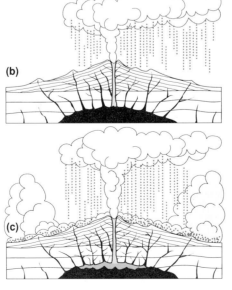

Topic Volcanic landscapes of the Pacific Northwest

The volcanoes of western Canada have been quiet for over 1000 years, although volcanoes are still very active in Alaska and in the Cascade Range from Washington State to California.

Our most noticeable legacy of volcanic activity is mountains such as Mt. Edziza on the Stikine Plateau, lava flows that cover most of the Cariboo and Chilcotin areas of central BC and hot springs such as Harrison, Radium, Fairmont, and Ainsworth. Many springs have become tourist spas. At Columns Provincial Park, northeast of Keremeos, a massive basaltic flow of the Tertiary Period (63-1 million years ago) is preserved.

Mt. Baker and Mt. Rainier are two of the biggest and youngest of the many andesitic cones in the Cascade Range. They appear to be dormant, but hot spots on their upper flanks and rims warn of continuing activity below.

Figure 1.22 Volcanic landscapes

- Earthquakes (5-8)
- Lava beds
- Hot springs
- Volcanoes

The dangers of volcanicity

Baked alive

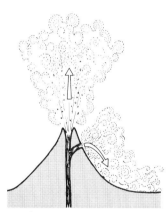

Figure 1.23 Volcanic disasters
The danger from a volcanic eruption depends on where the magma explodes into gas and ash. The Mt. Pelée disaster and Mt. St. Helens eruption occured because of high-level sideways explosions.

Although the mass of ash from the Helgafell volcano described earlier caused many houses to collapse, loss of life was avoided because there was time for evacuation to be effective. However, this is not always the case. When a natural catastrophe involves loss of life, it is termed a disaster. Without doubt one of the worst disasters happened to the people of St. Pierre on the Caribbean island of Martinique in 1902.

Mount Pelée, which overlooks the town of St. Pierre, began erupting on 23 April. Nobody was particularly alarmed by this because in 1792 and 1851 (the last occasions the volcano had erupted) the lava flows had been small. This time a small amount of ash was thrown up into the air and a small cone started to form. But at 7:50 on the morning of 8 May a series of violent explosions suddenly shook the mountain and a great black plume of ash rose many kilometres into the sky. Simultaneously another great cloud burst down the side of the mountain (Fig. 1.23). It was a spectacular sight, but few of those who saw it were to survive for long because the cloud hid a massive glowing avalanche of frothy lava, gas, and boiling liquid. Moving at 160 km/h the tongue of flaming gas ahead of the lava flow had raced down the mountainside and engulfed the town within two minutes leaving 30 000 people dead. Mount St. Helens erupted in the same way but the surrounding area was fortunately almost uninhabited.

The benefits of volcanicity

Volcanic gifts

On the Caribbean island of St. Vincent — near Martinique — a similar glowing avalanche swept down from the mountain top of La Soufrière, killing 1350 people. There have been many small eruptions of this volcano since 1902, including those of 1971, 1976, and 1979, but none caused much destruction. On all occasions many people were evacuated, just in case. With the volcano still erupting, this was the experience of one journalist in 1979:

❝ ... there were plantations of bananas and coconuts. It was near harvest time but the fruit trees had been left and, in any case, the bananas had been con-

Figure 1.24 Volcanic soil (a) These slopes of a volcano on the island of Java, Indonesia, are being used to produce rice despite the danger of a new eruption. (b) Plants can be grown directly in ash and lava only a short time after the eruption has ceased.

siderably damaged by the ash. The road, which was just a dirt track, was covered with coconuts. (Near the top of the cone) the terrain changed quickly. After climbing through grey ash for ten minutes the vegetation... disappeared completely — the gases from the eruption had burned it all... The climb was hard. It was difficult to find footholds and the further we climbed the worse it became. What some weeks ago had been a tropical forest was now a heap of ashes...

(Geographical Magazine)

As a result of these experiences you may wonder why people still live near volcanoes. However, a visit to the slopes of La Soufrière just a few years after any one of the eruptions would have shown the lava and ash weathered into a deep and fertile soil by the tropical rain and high temperatures. Equally quickly a luxuriant forest begins to reclothe the mountaintop in a rich green. And what suits the recolonization of trees, suits the growth of crops such as coconuts and bananas. The soils that develop on the volcano sides are the means of life on these islands and, in many ways, the people rely on periodic lava deposits and ash to renew the soils and provide fresh nutrients for the crops to use. The volcano is a blessing as well as a danger, and many people are prepared to brave the danger for the benefits (Fig. 1.24).

Hot springs and geysers

Hubble, bubble, toil, and trouble

When water moving in water-bearing rocks flows to great depths, it becomes hotter (Fig. 1.25) especially where molten magma is closer to the surface than is usual. In these areas the surrounding rock — and any water it contains — become very hot. Just as rock can be heated to great temperatures but remain solid if under pressure, when water is under pressure from the mass of overlying rock, it can be heated to temperatures well over 100°C. Sometimes there may also be faults which provide a direct escape route to the surface for the heated water. The hot springs and geysers that are major tourist attractions are often

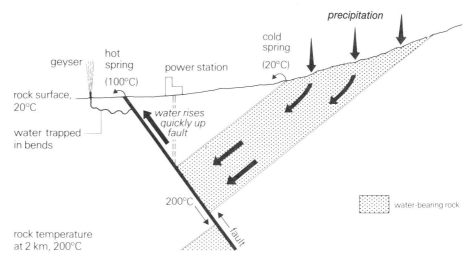

Figure 1.25 Hot springs and geyser formation
Water moves down through water-bearing rock. The fault provides an easy escape route for hot water which cools only slightly as it rises quickly.

at places where these faults reach the surface; for example "Bumpass Hell" (Fig. 1.26) and the 'Old Faithful' geyser in Yellowstone Park, USA, and "Geysir" in western Iceland from which the name geyser comes. At a hot spring, water bubbles to the surface constantly. By contrast a geyser forms when the hot water is temporarily trapped below ground in fissures that are often full of cold water near the surface. The build-up of pressure eventually becomes sufficient to eject both hot and cold water as a jet that may rise tens of metres into the air. After this there is a period of inactivity while sufficient pressure builds up again to cause a new spurt of boiling water. Mudpools usually occur where the supply of water is too restricted to produce hot springs. In these grey pools of boiling mud, gas rises continually and forms "chewing gum" bubbles which grow and burst every few seconds.

These features show that magma is still quite close to the surface. For example, "Bumpass Hell" in the Cascades (Fig. 1.26) is a region of hot water near a crater that poured out lava only 75 years ago. In this region, mud spatters and steam belches out of holes in the ground or rushes up into the air with a pulsating motion. At the same time continual release of sulphur gases gives an overpowering rotten-eggs smell. Together, steam and sulphur gas make breathing difficult within the hot springs basin.

Figure 1.26 Bumpass Hell
Geysers, boiling water and mudpools show that magma is still quite close to the surface. Good examples are found at Bumpass Hell, an aptly named crater on a mountain in the Cascades range of western United States. This geothermal area was discovered by K.V. Bumpass, a wilderness guide, in 1864. In this crater, steam belches out of holes in the ground or rushes up into the air with a pulsating motion. At Bumpass Hell water is provided in the spring by melting snow which seeps down cracks in the crater rocks. After the snow has melted away each summer there is less water available and many hot springs turn into mudpools. In these grey pools of boiling mud, gas continuously rises and forms "chewing gum" bubbles in the mud. These grow and burst every few seconds, splattering the mud out to the sides of the pools.

Topic The origins of mineral ores

Today, minerals are a vital part of any economy, so it is very important to understand how they may have been formed. Four of the most significant ways in which minerals accumulate and some of their most common minerals are: *1 Magmatic activity* may concentrate in magnetic iron ore, gold, platinum or diamonds. *2 Hydrothermal activity* may cause deposits of iron, copper, lead, zinc, silver, platinum, or uranium, usually as sulphides in fractured rock. *3 Sedimentary processes* have formed our greatest deposits of iron and evaporites (p. 31). These processes have also deposited gold, diamonds, platinum, and tin. *4 Metamorphic processes* concentrate minerals such as graphite and asbestos.

The Sullivan Mine at Kimberley, BC exploits a major iron-lead-zinc-silver deposit which is up to 100 m in thickness. About 1450 million years ago, sulphide minerals were deposited where hot, metal-bearing fluids rose through channels in the existing sedimentary pile to an ancient seabed. Later tectonic activity is the primary cause of the ore body's deformation. Movements associated with the westward drift of the North American continent during the past 150 million years also created massive folds and faults in what is now BC and Alberta. The Sullivan ore body is located west of the Rocky Mountain Trench, on the eastern edge of a north-plunging anticline (upward fold). The ore is extracted through a network of tunnels which have been dug in the country rock which is composed of mainly older, fragmented sediments. During the past 70 years, over 140 million t of ore have been extracted. The average mineral content has been 6.6% lead, 5.6% zinc, and 65 g/t silver.

Figure 1.27 Cross section of the Sullivan Mine, Kimberley, BC

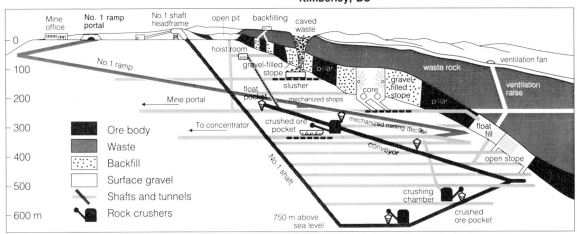

Exercise 1.4 Hot springs and geothermal power

1 Figure 1.25 shows how a hot spring forms. Rainwater moves slowly downwards through cracks in the rock, heating up as it goes. If the water moved more slowly to the surface would it still emerge hot? What is the importance of having a quick return route such as a fault (a long, continuous crack)?

2 Hot water provides a useful source of power in some areas where the water temperature is high enough (e.g. parts of Italy, New Zealand, and Iceland). When the water is to be used to generate power, a large-diameter hole is drilled to cut the fault at depth. What are the advantages of bypassing the fault and bringing the water to the surface quickly?

3 When the hot water reaches the surface it is no longer under pressure and it boils off as steam. The steam is used to drive turbines in the same way as conventional power stations. This type of energy is called **geothermal energy.** Why might geothermal energy be of significant value in meeting western Canada's energy needs?

4 There are hot springs in BC at Radium and Fairmont. In this region all of the conditions described exist except the water does not circulate very deeply. As a result, the water reaches a maximum temperature of 80°C at depth, and by the time it reaches the surface it's 46°C. Explain why it is impossible to use the hot springs in the Invermere area to generate electricity.

2 Rocks from sediments

Sedimentary rocks

New land for old!

Originally, the earth's crust consisted only of igneous rocks made from cooling magma. However, rain and running water soon broke down the igneous rocks into small particles and carried them away to the sea. In this way material for the first **sedimentary rocks** was formed. Sedimentary rocks (mainly those made from the eroded remains of earlier rocks) now make up a major part of the surface rocks of continents, and igneous rock exposures are less extensive. Because most sediment is brought to the sea by rivers, it builds up in the shallow seas near continents. There is only a thin layer of sediment in the deeper parts of the ocean basins, and igneous rocks still dominate the ocean crust.

The way in which sedimentary rocks form and build up into layers is a very important influence on landscape development. Variations in the compaction of sediment and the types of rock debris produce rocks that show much variation in hardness and resistance to erosion, characteristics which are seen very clearly in the Grand Canyon (Arizona, USA) because here the rocks have been exposed by weathering, mass wasting (rock fall of weathered rocks), and erosion.

The Grand Canyon story

Back to base

In southwestern Arizona you can walk a few steps from a modern road and find yourself standing on the edge of the greatest valley in the world (Fig. 2.1). This is the Grand Canyon, which has been cut through 1000 million years of earth's history to reach down a breathtaking 2 km to the Colorado River below.

Figure 2.1 The Grand Canyon (Arizona) This canyon has been eroded through over 2 km of horizontally bedded sedimentary rock.

One of the most impressive features of the Grand Canyon is its many-coloured walls which descend from the rim like a giant staircase. The "treads" of the staircase have been formed as each rock layer has been eroded back by a different amount, depending on its hardness.

From the rim of the canyon steep hiking trails snake downwards. Following one such trail, at first you cross over a 100 m layer of white limestone — a mere 200 million years old — then comes 400 m of coarse deep-red sandstone. This in turn is followed by more limestone and then thick beds of shale. The final gorge reaches down through yet more layers of rock, finally exposing the top of a granite batholith.

All of the rock layers above the granite are made from the eroded remains of earlier rocks; that is, they are sedimentary rocks. In the Grand Canyon the rocks are piled up horizontally, one upon the other, with the youngest rocks on top and the oldest ones below. The upper ones contain **fossils** — the remains of animals and plants that have been turned into rock and preserved with the other sediments. Fossils are particularly useful because they tell us of the environment that existed when the rock was being laid down. For example, fossils in the limestone near the rim show that the Grand Canyon was once part of the seabed, even though today it is more than 2 km above sea level!

Here, without doubt, is proof that sedimentary rocks are part of a cycle of events that takes eroded material out to the oceans, forms it into new rocks, and then lifts it up into land once more.

The Grand Canyon is spectacular partly because we can still see the rocks in the horizontal layers as they were first laid down. In many other parts of the world the same types of rocks have been laid down, but they are not as easily seen. For example, a cross section through the sedimentary materials below the Canadian Prairies clearly shows a build-up of rocks far greater than in the Grand Canyon (Fig. 2.2). In the foothills of the Rockies and against the Canadian Shield, the edges of these ancient sediments have been weathered and eroded, making the geological sequence difficult to read.

Today sedimentary rocks on land are far more widespread than igneous rocks, even though igneous rocks may dominate the landscape locally. The cycle of erosion from the land, deposition in the sea, and compaction into new rock is now the main process of rock formation.

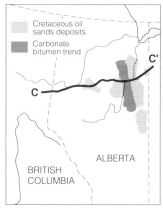

Figure 2.2 A generalized cross section of the northern Canadian Prairies
Sediments, which are over 5000 m deep, in the Rocky Mountain foothills vary from recent glacial till on the surface to granite wash and Cambrian basement rock at greater depth.

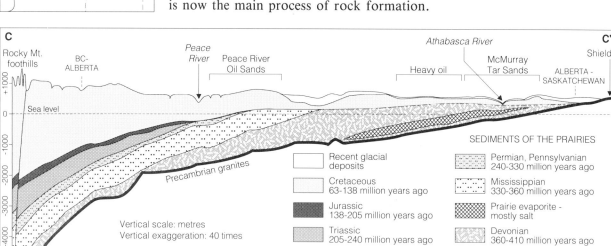

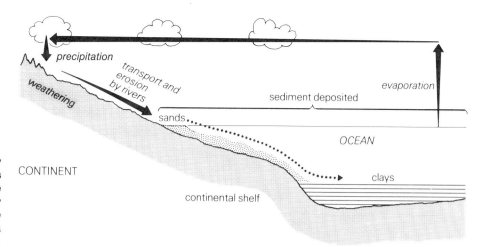

Figure 2.3 Sedimentary rocks
Sedimentary rocks form as the sediments are compacted by the mass of new debris. The shells and skeletons of sea animals are also present.

Sedimentary processes

From rivers to rocks

Sedimentary rocks are made in many ways. Mostly rivers carry eroded debris down to the sea where it settles out in the relatively still water. Coarse particles such as sand settle out first near the shore and eventually become compacted into sandstone; fine materials such as clay settle out in deeper water and eventually become compacted into shale (Fig. 2.3). However, in the past other forms of transport have been important (Fig. 2.4). Glaciers, volcanoes, wind, and coastal waves have all played their part in carrying materials away from the land and out to sea. Remember that sediment builds up on the seabed very slowly — perhaps less than a millimetre a year. In the past, climates have changed considerably and the sea floor has been moved up and down many times. This is why very different rocks have built up one upon another. Figure 2.2 shows that, although sandstones and shales are being formed in the Prairies today, in the geologically distant past sediments have included materials of tropical origin such as limestones, coal, potash, and salt (p. 31). These sediments, which contain oil and gas, are 200-400 million years old.

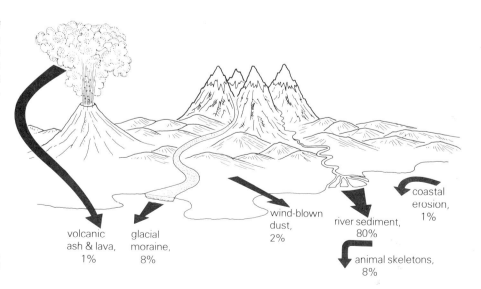

Figure 2.4 The sources of sediments in the oceans
The proportions shown here are based on recent estimates; in the past these proportions have varied considerably.

Coral formation

Rocks that live and grow

Although many sedimentary rocks are made of tiny particles of rock that settle on the sea floor, some rocks are made almost entirely from the shells of sea creatures. These creatures produce their skeletons or shells by extracting dissolved materials from sea water. The rocks they form are called **limestones.** The remains of shells are easy to see in some limestones although in rocks such as chalk the skeletons are so small that you would need a microscope to identify them.

Coral reefs are formed by the slow accumulation of countless numbers of small, interconnected shells of which only the outer layer is alive. Most marine creatures remove some calcium from sea water and eventually add their skeletons or shells to oceanic sediments.

Warm seas often provide the best conditions for marine creatures to grow, because food supplies are abundant. But the shells of the creatures can still be lost if a large amount of sand or clay is also settling on the sea floor. So even in warm seas limestones are still uncommon, and in general shells make up no more than the occasional fossil in most rocks.

Because of these restrictions shells build up into limestones most effectively when there is little sediment produced on the land. Suitable conditions occur off the coasts of low-lying desert regions, or on remote islands in the middle of oceans.

Coral reefs are made of limestone (Fig. 2.5). Often they grow around islands or in the form of rings called atolls (Fig. 2.6) but their extent is very

Figure 2.5 Fringing reefs
This fringing reef has formed around an aging volcano. It shows as the light blue band in the azure blue sea.

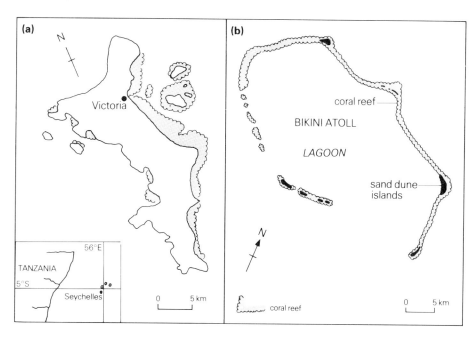

Figure 2.6 Coral reefs (a) Mahé, Seychelles and (b) Bikini atoll.

limited. Many limestones do not contain any coral at all. But whatever the source of the limestone material, it is soon formed into rock by burial under other sediments.

Some limestones are forming today. For example, coral reefs occur off the northeast coast of Australia (the Great Barrier Reef, Fig. 2.7), while finer-grained limestones are building up off the Florida coast of the US. However, in the past climates have been warmer and limestones have been much more extensive. Limestones have formed and been weathered into distinctive landscapes in the Karst Plateau of Yugoslavia, the White Cliffs of Dover, and the Dales of northern England (Fig. 15.6); New Zealand's volcanic plateau; and throughout much of China, especially in Kwangsi Province. The best-known North American examples are in Puerto Rico, Cuba, and the Yucatan Peninsula; central Florida, Kentucky, New Mexico, and Yellowstone Park; the Niagara escarpment, the Canadian Rockies, and Vancouver Island. Chapter 15 describes the special features of these areas which are attractive for the tourism industry and also cement manufacturers.

Figure 2.7 The Great Barrier Reef, Australia

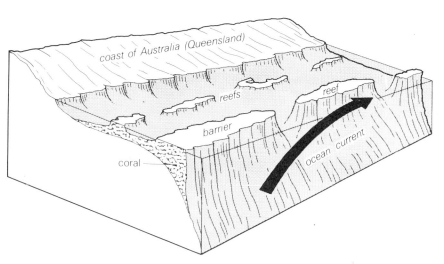

Topic Coral reefs and atolls

Most reef-building corals live in warm, clear water with plenty of sunlight. They grow fastest in the top 10 m and rarely grow in water deeper than 70 m. A coral reef is made only partly from coral animal skeletons. A large part of the reef is made from tiny plants called algae that help fill in the spaces between corals.

All coral animals need a constant food supply and, because they are fixed to the reef, they take the food that drifts past them in ocean currents. Consequently, coral reefs grow best on their seaward margins; they often cease to grow on the side of the reef facing the land. The **lagoons** between the reef and land never fill with coral because there is little food in water cut off from ocean currents.

Coral reefs are found in several distinct forms:

1. *On continental margins.* Here reefs are very extensive, e.g. the Great Barrier Reef, Queensland, Australia (Fig. 2.7) which is 2000 km long.
2. *On oceanic islands.* Here reefs occur in much smaller, isolated forms. They are three main subgroups: (a) **fringing reefs** (e.g. Mahé, Seychelles in the Indian Ocean, Fig. 2.6) where the reef grows right up to the beach, (b) **barrier reefs** where the volcanoes around which they form have been sinking for millions of years, and (c) **atolls** (e.g. Bikini, Marshall Islands, Pacific Ocean, Fig. 2.6) where the volcanic island has sunk out of sight and a lagoon is left within the ring of coral. The small atoll islands are made from broken coral "sand" that forms dunes 3-5 m above sea level. Even coconut palms have colonized the larger dune islands, making them classic "tropical paradise" islands. The sequence of development of these islands is shown in Figure 2.8.

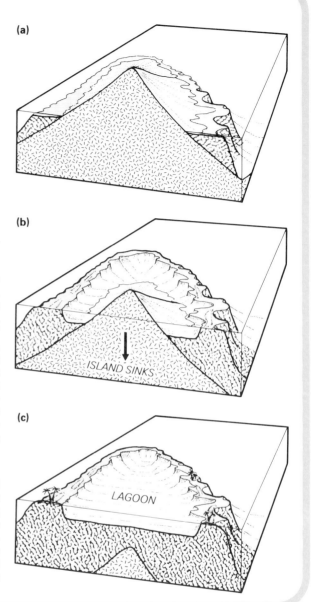

Figure 2.8 Types of reef
(a) With a reasonably constant sea level a **fringing reef** develops. (b) If sea level rises or the land sinks, the coral has to grow upwards to keep within the zone of sunlight and plankton food supply. A channel develops behind the new **barrier reef** because coral cannot grow away from the open-sea food supply. (c) An **atoll** is usually a volcanic island which used to have a fringing reef; but as sea floor spreading and the mass of the volcano caused it to sink, the coral grew upwards, enclosing a lagoon.

Oceanic minerals ## Oceans of plenty

In oceans, eroded materials finally come to rest. We might be tempted to call them nature's "garbage dumps" and yet, along with the sand and clay, millions of tonnes of valuable minerals are continuously poured in. In many cases these materials are so scattered that it is too expensive to recover them, but occasionally natural processes concentrate these resources. So when we search for minerals we use our knowledge of nature to find places where natural concentration has occurred. Sediments often contain large deposits of minerals. Possibly the most important minerals for industry are gas, oil, coal, iron, and salt. Coal in particular has influenced the location of much of the world's industry.

The sediments which lie deep along continental margins in areas such as the Persian Gulf, North Sea, Grand Banks, Gulf of Mexico, Gulf of Maracaibo (Venezuela), the Los Angeles-San Diego Basin, Bass Strait (between Australia and Tasmania), and Indonesia and Malaysia contain many valuable minerals. In all of these areas, the accumulation of sediments has occurred for over 300 million years. Oil and natural gas are the most accessible of the submarine sedimentary resources. Today it would be difficult to overestimate the importance of oil and gas in our daily lives, not just for transportation, heating, and lighting, but also in the manufacture of pharmaceutical drugs, building materials, and synthetic fibres. The rocks and minerals which occur below sea level, as part of continental shelves, are rarely very different from the sediments that cover the nearby continental areas. However, with few exceptions the extraction of oil and gas from land areas closest to bodies of water has predated offshore rigs by several decades.

In Western Europe, the sedimentary layers that contain North Sea oil and gas also contain coal seams that outcrop in Britain and Germany — coal that fuelled the Industrial Revolution.

Building materials ## At home with rocks

Igneous, sedimentary, and metamorphic rocks provide the raw material from which the landscape has been fashioned. These rocks are also basic to our lives, our structures, and our cultural environments. We use rocks to build roads and rail beds, walls, bridges, buildings, and houses.

The pyramids of Egypt, the Acropolis of Athens, the Great Wall of China, and the castles and cathedrals of Western Europe all attest to the importance of stone in building throughout our history. However, stone quarrying does not lend itself to economical mass-production of building materials. Today most houses use natural stone for decorative purposes only, such as fieldstone fireplaces, slate hearths, and white quartz for landscaping. But modern houses owe more to geology than meets the eye. "Gypsum" wallboard is mass-produced from gypsum (calcium sulphate) and processed into easy-to-use panels that line most of our homes, schools, and offices. Fired-clay bricks form the outer walls of many public buildings, factories, and some homes. But such bricks can be made from only certain types of clay, so they have been largely replaced by a concrete alternative — a mixture of sand and cement. These concrete blocks are used to build many of our schools and low-rise buildings, such as shopping centres, halls, and arenas. Concrete blocks are also used to make the coloured paving bricks that decorate our sidewalks, patios, and driveways. Sand and gravel are combined with powdered, baked limestone (portland cement) to form concrete which, in its various forms and applications, is indispensable to modern construction. We even mine volcanic cinders to provide rocks for our barbeques and saunas!

Massive slabs of highly polished granites and marbles are still used to build and decorate some of our most imposing public and corporate buildings. But, for the most part, we have rearranged our geological resources in such a way that we take them for granted.

Topic   The basis of industry — coal, gas, oil, iron, and evaporites

Coal

Over 200 million years ago coal deposits were formed from dense forests of large trees. These trees grew in tropical swamps and as they died they fell into the stagnant swamp water and thus, without air, were protected from rotting. However, from time to time the land sank quite quickly, the sea flooded in and it covered the fallen trees with sand and mud. Eventually the land rose, and new trees grew on the sand and mud until there was a new period of sinking. Mostly there was only enough time for thin tree layers to build up between deposits of sand and mud. As a result of the mass of the overlying sediment, the coal seams were squashed much thinner than the original tree layers. Today it is the thin seams scattered among sandstone and mudstone that the coal miners seek, so it is no surprise to find that coal is often mined at a number of different levels.

Because coal was formed a long time ago there has been plenty of time for it to be buried under hundreds of metres of rock and to be affected by later movements of the earth. Two particularly important earth movements slowly formed the Appalachians and Rocky Mountain systems about 240 and 140 million years ago, respectively. Much of the material which originally formed the mountains has been eroded, exposing some of the remaining coal while leaving other coal beds deeply buried. In Atlantic Canada, coal is mined by underground methods at the Sydney-Glace Bay area of Nova Scotia. In the Rockies, strip mining is the most cost-effective technique, although the most destructive to the environment. Extensive shallow deposits of sub-bituminous coal and lignite (medium and low-grade coal) are accessible by surface mining throughout most of southern Alberta and the southern extremity of Saskatchewan.

Gas

Coal provides us with a fuel to burn. Also during its compaction from peat, it gave off gas. In many regions the gas has subsequently risen through the overlying rocks, reached the surface, and been lost. However, some rocks, especially salt beds, are "gas tight" and when gas rises from beneath them it becomes trapped. Despite past losses, therefore, huge volumes of gas are still trapped in the rocks above coal fields, particularly under the Canadian Prairies and North Sea where such coal is too deep to be mined. The gas has risen and been trapped in upward folds of the rocks, so that it now occurs in isolated pockets. Each of these pockets of gas has to be found and then a well drilled before the gas can be piped to refineries. It is an expensive and difficult process but the rewards are huge.

Figure 2.9   The distribution of coal fields and oil and gas fields and pipelines in Canada.

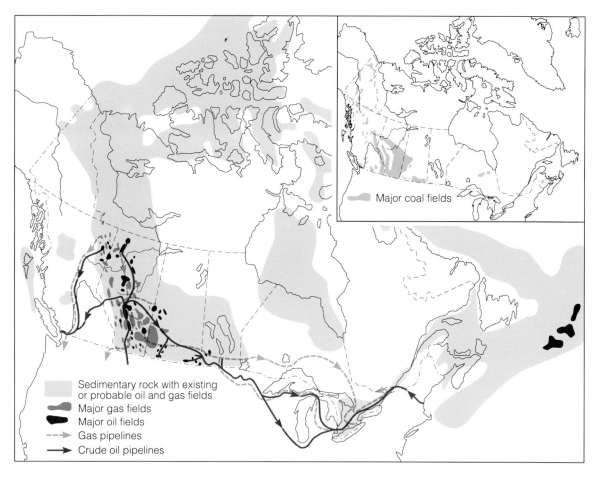

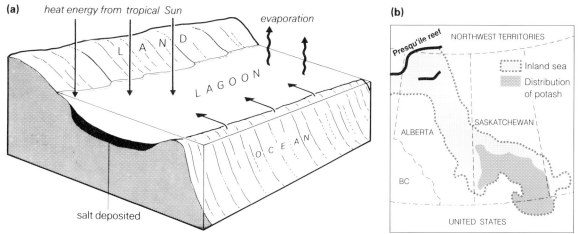

Figure 2.10 Evaporite formation
(a) Evaporites are formed in tropical coastal lagoons or inland lakes. (b) Such lagoons once formed in what is now the southern half of Saskatchewan.

Oil

Oil is formed by the slow decay of the bodies of microscopic sea creatures that were buried in the muds on a sea floor. Although these muds contain only a tiny proportion of dead bodies, their slow decay releases oil and gas which then seeps upwards through the rocks above. It is trapped below upward folds of salt beds or mudstones just like the gas from coal. Some of these oil and gas pockets lie under the Canadian Prairies at depths ranging from 250 m to 7000 m, and beneath the Grand Banks of Newfoundland where deep water, icebergs, and storms increase the risks and costs of exploration (Fig. 2.9).

Oil and gas are found in many parts of the world. The largest reserves are in the Middle East where extraction is easy. There are other large oil fields in Texas, Venezuela, Mexico, and Central USSR. There are also large reserves of oil in places such as Arctic Canada and Alaska. Oil locations have forced people to work in inhospitable climates. After the initial drilling pipelines were built so that the oil could be taken to the main population centres. Being more easily transported than coal, oil and gas travel to the cities in contrast with the factories and settlements built on the coal fields in 18th and 19th century Europe.

Iron ore

Some iron ore is found near to magma areas as described on page 22. The "mountain of iron ore" at Kiruna in northern Sweden is probably an example of molten iron cooling out into a very rich vein. Iron ore at Kiruna is 68% iron and is so rich that it can be mined profitably even in such a remote location.

Most iron ore is of considerably lower quality. In Europe the large ore fields contain less than 30% iron. Nevertheless these ore fields are another example of the way natural processes concentrate minerals in the sea. Rich ore veins like Kiruna have been eroded from mountains and then transported to the sea in solution. However, sometimes this eroded ore can be concentrated in sedimentary rocks. Limestones and sandstones often contain precipitated iron in sufficient concentrations for mining to be worthwhile. Occasionally bands of iron-rich sandstones also occur intermixed with coal seams; this is called "black band" iron ore. Most of Britain's early industrial areas were based on these chance occurrences of two raw materials together. The rich iron ore deposits of the Labrador geosyncline (depression) and to the west of Lake Superior provide reserves of at least 20 000 million t. Labrador produces almost 50% of Canada's iron ore.

Evaporites

Our last examples of raw materials for industry are evaporites which precipitate from sea water. Salt is formed in coastal lagoons. The sun's heat causes the lagoon water to evaporate, concentrating the salt water until some salt is precipitated on the lagoon bottom (Fig. 2.10). New sea water continually flows from the sea to replace the water evaporated and this is evaporated in its turn. By this process, salt beds hundreds of metres thick have been built up only to be buried by later sediment. As sea water, which contains about 3.5% "salt," evaporates, 3 different salts are precipitated. As water becomes increasingly more saline, gypsum (calcium sulphate), then halite (sodium chloride), and lastly potash (potassium chloride) are deposited. Saskatchewan potash was evaporated from a long, narrow inland sea about 360 to 380 million years ago. This sea became a vast salt pan, and sea water was occasionally added through the Presqu'ile Barrier Reef from the northwest. Evaporation progressively freed the salts from solution until, at the southeastern end of the sea, potash was laid down in 3 main layers of 4.5 m to 6 m thickness. Limestone, shale, halite, and clay are mixed with the potash, so the ore must be processed before use as fertilizer.

Salt is also deposited in inland lakes in desert regions. The Great Salt Lake and the Bonneville salt flats, Utah, USA; and the Dead Sea on the Israel/Jordan border are examples of land-locked lakes where river water is continually evaporating and leaving its salt content behind. These areas, and many more, provide a valuable resource wherever they occur. For example, both Israel and Jordan have recently built chemical plants beside the Dead Sea.

Rock, when broken into small pieces of gravel and sand, is widely used for road and railway foundations. Gravel provides an excellent base because it is highly permeable, so it does not hold water which would cause frost heaving in cold weather. Throughout most of Canada, gravel is abundant and readily available in ancient **moraines** and **outwash plains.**

Finally, limestone can be processed to make agricultural lime which helps reduce the acidity of soils; and it can be baked to make portland and mortar cements. As a result, many cement and fertilizer plants are located very close to outcrops of limestone.

Exercise 2.1 Coal mining: Quinsam Coal Mine

Near Campbell River on Vancouver Island, layers of clean, highly volatile coal are being mined within 25 km of tidewater. Great care has been taken to protect the environment from pollution during and after each stage of mining and moving the coal. The mining company has guaranteed the maintenance of good water quality and the protection of delicate and invaluable fish habitats in local streams. An active environmental protection department is maintained, and a public information office is kept open at Campbell River.

The open-pit mine will yield up to 30 million t of coal at the rate of as much as 1 million t a year. The mined land is being reclaimed as high-quality coniferous forest.

1. In what type of rock were the Quinsam coal deposits formed?
2. The upper coal seam is 4.5 m thick and the lower seam averages 3.5 m in thickness. At position A in Figure 2.11, what is (a) the total thickness of the coal seams? (b) the depth of soil and rock above the lower seam?

Exercise 2.2 Problems related to oil and gas exploration

1. Figure 2.9 shows the distribution of Canada's coal deposits and sedimentary rocks in which oil and gas fields exist or might exist. There are, in fact, hundreds of small oil and gas fields rather than one or two big ones. Why is this so?
2. A recent development in the search for oil and gas has been offshore exploration. In small groups, brainstorm a list of probable problems and possible solutions associated with oil and gas exploration and exploitation on the Grand Banks.
3. Oil and gas fields are not close to areas of high population and industrial concentration. Research the exploitation, transportation, and refining of oil and gas. Write a 1-page report summarizing the main problems and attempted solutions.
4. How can the problems you have listed in #3 be resolved in a cost-efficient manner?
5. Because oil and gas occur in small pockets, do you think the chances of finding more pockets is high or low? Why?

Figure 2.11 Cross section of Quinsam Coal Project

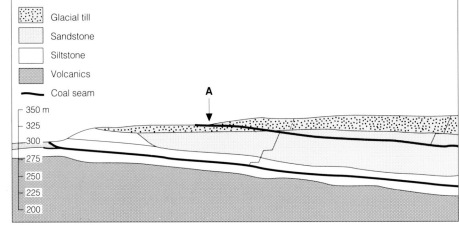

3 Earthquakes, faults, and folds

Faulting and earthquakes

All fall down

Figure 3.1 Earthquakes The spread of shock waves from the disaster area of the Armenian earthquake of 1988.

Twenty-five thousand people died in an earthquake that devastated two cities and wiped out an entire town in the southern Soviet republic of Armenia yesterday.

In the region of Spitak, which had a population of 55 000, almost no one survived the 7 December 1988 earthquake. Three quarters of Leninakan, Armenia's second largest city, was destroyed.

Such a newspaper report gives little indication of the real impact of the event (Fig. 3.1). In Spitak, the shock of the 6.9 magnitude earthquake, which struck at midday, was so great that most buildings disintegrated. Roofs crashed down, apartments were flattened, and thousands of homes were reduced to rubble. During the earthquake, Armenians huddled beneath what shelter they could find and hoped they wouldn't be crushed by the debris. Throughout the next two weeks, survivors — many of them severely injured — were rescued and taken to the already over-crowded hospitals in nearby cities.

Earthquakes occur deep in the rigid crust of the earth. We see the results of ancient earthquakes every time we look at a section of cliff and notice some rocks are displaced with respect to others (Fig. 3.2). A **fault** (displacement can

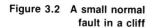

Figure 3.2 A small normal fault in a cliff

Figure 3.3 The Great Glen Fault
Erosion has etched out the fault into a trough that cuts right across Scotland. Faults often cause a rock to shatter along the line of movement and this allows erosion to proceed more quickly than elsewhere. Note there has been no vertical displacement.

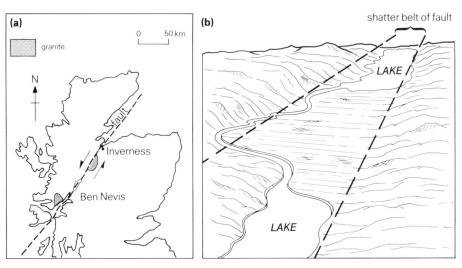

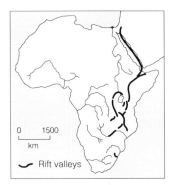

Figure 3.4 The great rifts of East Africa
Massive valleys caused by normal faults extend from the Dead Sea to South Africa.

Figure 3.5 The San Andreas fault
This tear fault still moves about 5 cm/a on average. The epicentres of two 1986 quakes are marked.

represent the movement of millions of tonnes of rock and give us some idea of the size of the forces that strain and tear at the earth's crust. Even a single fault at depth sends many shock waves to the surface causing an earthquake to last for several minutes. You can compare this with throwing a pebble into a pond: the pebble strikes the water and is gone, but the ripples it creates wash the shore of the pond for quite a long time afterwards.

Some faults are of a stupendous size. For instance, subsurface forces in the past slowly tore Scotland in two (Fig. 3.3). As the two halves of Scotland scraped past one another for tens of millions of years, a whole belt of rock was crushed. Erosion has now exploited the weakness of this crushed rock, leaving a gash so big that it takes three lakes just to cover the bottom. In Africa a complex series of trenches extends 6400 km from the Dead Sea through the Red Sea to Uganda and Kenya to South Africa. These great valleys are 32 to 128 km wide (Fig. 3.4).

Everyone in California, USA, knows about earthquakes and faulting because San Francisco is built across the most famous fault in the world. This, the San Andreas fault (Fig. 3.5), is the same type as the one which tore Scotland in two. The fault line slices through the western coast of the state, showing the place where a small strip of coastal California is moving slowly northwards against the bulk of the American continent in a series of jerks. Although this has been taking place for millions of years, very few months go by without a new movement. Most are very small and only give minor shock waves (tremors). The last time the San Andreas fault had a major movement was in 1906 and then much of central San Francisco was destroyed. Some people predict that another major earthquake will strike San Francisco within the next few years, but as yet there is no accurate way of forecasting the shock.

As the San Francisco experience shows, faults move in jerks. It may take tens or hundreds of years before forces build up sufficiently within the crust to break slabs of rock, but when the break comes the rock snaps into its new position within seconds or minutes.

Earthquakes tell us which pieces of the earth's crust are most active. It is clearly not a coincidence that, in general, these are the places where volcanoes are most commonly found. This is because a large fault can easily provide a

Exercise 3.1 A major fault in Scotland

The northern part of Scotland once lay further northeast than it does now. A huge tear fault occurred, moving northern Scotland very slowly in a southwesterly direction over millions of years, shattering all the rocks along the sides of the fault in the process.

1 Copy the part of Scotland north of the fault on Figure 3.3a. Now copy the rest of Scotland, but displace it in such a way that the granite areas match. These granites were once part of the same batholith and they tell us how far the fault has torn. The map you have now drawn shows us what Scotland once looked like.
2 Measure and write down by how much the granite areas have been separated.
3 The tear fault shattered the rocks along its length so that they were weakened and more easily eroded. This zone has subsequently been eroded into a huge trench. Use your atlas to name the three lakes (lochs) that occupy the deepest parts of the trench.

route for magma to escape to the surface. In western Canada, earthquakes are most apparent along the Pacific Coast where sea floor spreading is still active. The frequency and intensity of earthquakes decreases quickly with distance eastward from the coast. Mountain building in the Rockies appears to have slowed recently, and the ancient Canadian (granitic) shield is very stable.

Although many rocks are very brittle and can yield to pressure only by faulting, many others have simply bent (Fig. 3.6). Bending or **folding** of rocks is a very slow process. It may occur in sediments that have been deposited on a sea floor only recently, or in deeply buried rocks if the pressure is applied over a wide area. Indeed, folding is the dominant process in most major mountain systems. Consequently they are usually called **fold mountains** (Fig 4.1).

Even rocks that are brittle when cold can fold when they become hot, deep underground. (Compare this with a bar of toffee which breaks when cold but bends when warm.) Folds in hot rocks probably occur deep down in the crust and they are usually associated with the formation of mountain ranges. We only see this type of folding when the mountains are eroded and the "cores" of the mountains are exposed at the surface.

Figure 3.6 Folded layers of sedimentary rocks

Folds are permanent deformations of rocks, but a rock can store up energy if bent only slightly in just the same way as a watch spring (Fig. 3.7). This energy can be released again if the bending pressure is removed. The energy is used to return the rock to its original position. For example, only a few thousand years ago huge ice sheets covered the northern continents. Their mass pressed down on the land, forcing it downwards. Now the ice has gone and the land is slowly returning to its original position. Some parts of Scotland are still rising by about 4 cm a century; parts of North America have not recovered as rapidly and parts of the depression remaining are still occupied by the Great Lakes.

Figure 3.7 For caption, see Summary opposite

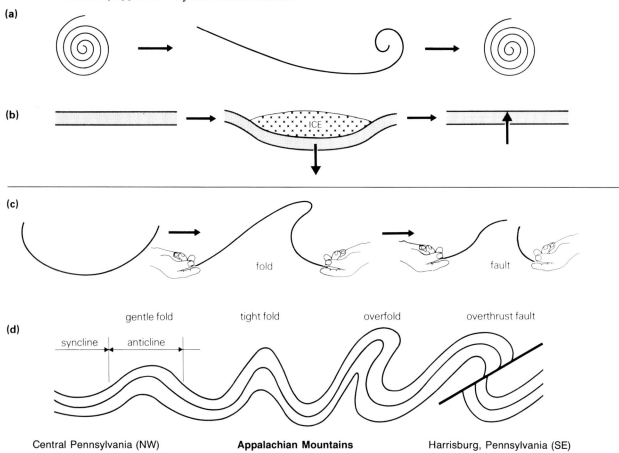

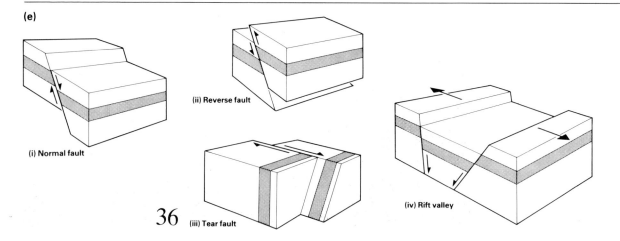

SUMMARY: ROCKS UNDER PRESSURE

If a rock is bent too far it will change its shape permanently. This produces a **fold** (Fig. 3.7c). If bent severely the rock may even break. This produces a **fault** (Fig. 3.7e).

An upward fold is called an **anticline**, a downward fold is called a **syncline**. The tightness of the fold is an indication of the bending pressure of forces in the earth's crust (Fig. 3.7d). As the pressure becomes greater the folds become tighter. When pressure is applied from one side more than the other the folds lean over and are then called **overfolds** (Fig. 3.7d). Gentle folds are often very broad. For example, the folds that have shaped central Pennsylvania are many tens of kilometres across.

In some cases rocks are too brittle, or the pressure applied is too great, for folds to develop, and then the rocks fault. Most faults move rocks in a more or less vertical direction (these are **normal** or **reversed faults**) (Fig. 3.7e, i & ii). It is rare for faults to occur singly and they are usually found in groups. Large faults can produce distinctive landforms such as **rift valleys**. (Fig. 3.7e, iv) (e.g. the Rhine rift valley) or uplifted block mountains known as **horsts** (e.g. the Mount Lofty Ranges, South Australia, Fig. 3.12) **Tear faults** occur when rocks scrape past one another with little vertical movement (e.g. the San Andreas fault) and can usually be seen only when erosion removes the shattered rocks alongside the fault line (Fig. 3.7e, iii).

Figure 3.7 Folds and faults (*opposite*)
(a) A coiled watch spring can be pulled out, but it will return to its original shape when released. (b) The loading of ice on continents bends the rock downwards temporarily, but it recovers after the ice melts. (c) If a spring is bent too far, it changes shape permanently and may finally break; the same applies to deformed rocks. (d) Types of folding. (e) Types of faulting.

Exercise 3.2   Rift valleys

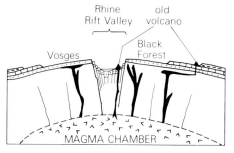

Figure 3.8 The Rhine rift valley

Figure 3.9 El Capitan in the Yosemite Valley (California) Notice the unbroken surface of this rock. The absence of joints is the main reason for its resistance to erosion.

1. Study the diagram of the Rhine rift valley and adjacent areas. (Fig. 3.8). What information tells you this is a rift valley?

2. Find a map of East Africa in your atlas. Locate Lake Victoria at 2°S 33°E. To either side of this lake are long narrow lakes that occupy the two arms of the East African rift valley. In fact the rift system starts in the Middle East near the Dead Sea on the Israel/Jordan border (32°N 36°E).

3. Make a sketch of East Africa and the Middle East, and mark on the rift system from the Dead Sea to Malawi (14°S 35°E). The East African rift system is still active and many small earthquakes occur frequently.

Exercise 3.3   The Yosemite Valley — a landscape guided by joints and faults

El Capitan means "the chief," and it is the name given to the tallest unbroken cliff in the world. El Capitan rises vertically 1000 m above the floor of the Yosemite Valley in California, USA (Fig. 3.9). The most experienced climbers in the world come to pit their skills against this famous slab of granite and it still takes them three days to reach the top. They need a mass of special climbing gear, as natural footholds are so few. This is because the El Capitan granite has not developed many joints since its formation.

And just as climbers rely on joints for gaining climbing foot or hand holds, so nature uses joints for increasing the rate of erosion (see p. 112). Where joints are few, so the effects of rain, river, and ice are slow. The Yosemite Valley is at its narrowest where the El Capitan granite is found. But just a little further along the valley the rocks change, the number of joints increases and the valley widens out. A closer inspection of the valley shows that whole cliffs are guided by "master" fractures in the rock. For example the "Three Brothers" cliff shows how fractures in three directions have been exploited to erode the rock into this unusual shape (Fig. 3.10).

1. Draw a sketch of the Three Brothers (Fig. 3.10) and mark on the other two directions of master joints that have determined the shape of this cliff.
2. Valley shape and the course of tributaries are also controlled by master fractures. Look at the map (Fig. 3.11). There are two main lines of fracture at right angles. The Nevada Falls flows across one trend line, the Vernal Falls flows across the other. Draw the valley and mark in sections that run parallel to the master fracture directions. This shows you the extent to which faults and fractures can influence the shape of a landscape.

Figure 3.10 The Three Brothers, Yosemite Valley
Joint control has determined their shape, the black line shows one direction of "master" joints.

Figure 3.11 Landscapes influenced by joints and faults
The Yosemite Valley has been eroded by water and ice, but the shape is still largely due to its pattern of joints and faults.

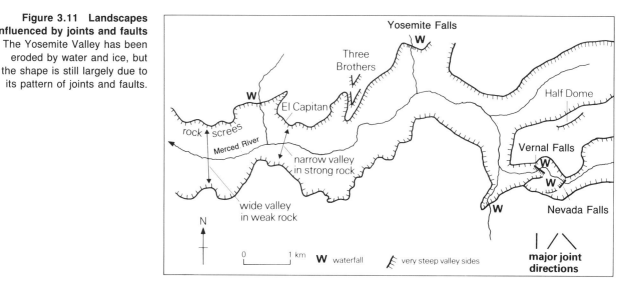

Figure 3.12 The shatter belt of South Australia
A series of 7 major normal faults has caused 2 rift valleys which have been flooded by the sea. The Mount Lofty Horst rises 734 m above sea level.

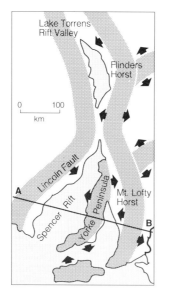

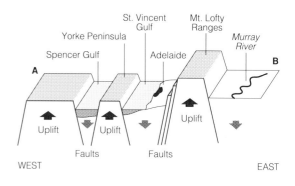

Figure 3.13
The Rocky Mountain Trench extends 1600 km in a virtually straight northwest-southeast line in BC. It is very narrow (3.2 - 16 km) and generally flat-bottomed. The trench's two major lakes, Williston and McNaughton, are both reservoirs of large dams.

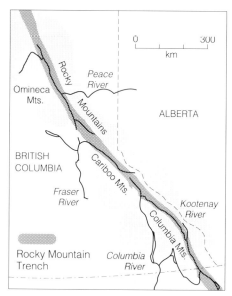

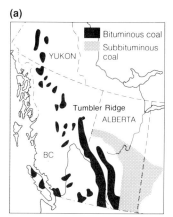

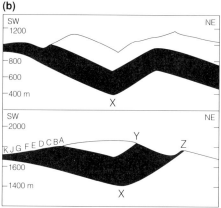

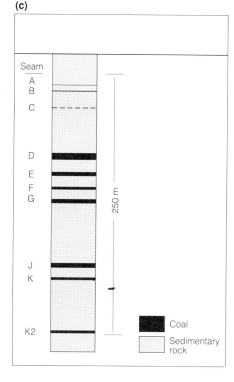

Figure 3.14 Coal fields
(a) location of coal deposits in British Columbia
(b) cross sections of 2 synclines which are being strip mined
(c) profile of the 9 main coal seams

Exercise 3.4   Quintette Coal

1. What are the features marked x?
2. What is the distance from y to z?
3. In the coal-bearing deposits, what is the ratio of coal to rock?
4. What technique is used to mine this coal?
5. Study a map of BC to determine the best route to export the coal.

In northeastern British Columbia, near Tumbler Ridge, 2.9 billion t of coal has been found in the folded, faulted foothills of the Rockies. The coal-bearing strata are mostly deltaic deposits dating back about 200 million years. The main coal seams are in broad synclines and sharper anticlines. Thrust faults and erosion have exposed the coal at the surface. A total of nine coal seams have been identified. Their total thickness, on average, is 19 m.

Four open-pit or strip mines have been developed. They are designed to operate through the next century.

4 Continental drift and mountain building

Continental drift

The moving shell

In the previous chapters we have seen that volcanoes and earthquakes are the signs of major lines of weakness in the earth's crust. We have also seen that these weaknesses make up clear patterns, one of the most striking of which is the Pacific Ring of Fire (Fig. 1.5). These lines of weakness really split up the crust into large slabs, usually called **plates** (Fig. 4.1). Once again, think of the crust as being rather like the shell of a hard-boiled egg that has been cracked with a spoon before being eaten. Jagged edges appear everywhere over this thin and brittle shell, yet individual pieces remain intact; the intact pieces of rigid shell are plates and the jagged edges are called **plate boundaries.** It is along the plate boundaries that we find volcanoes and earthquakes.

On page 7, we noticed that the temperature contrasts between the hot core and the cold crust probably causes convection currents to creep over the rocks of the earth's interior. These currents would cause a massive drag on the earth's crust causing whole pieces of it (the plates) to move. However, every part of the earth's surface is occupied by crust. This means that each time a plate moves it must collide with, scrape past, and split away from its neighbours.

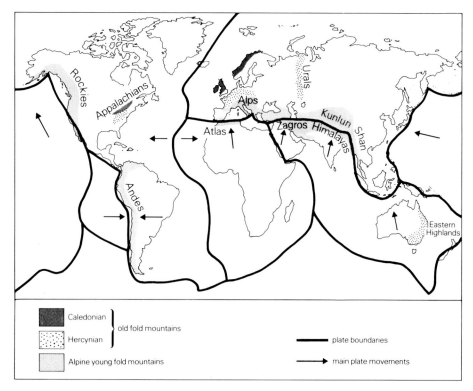

Figure 4.1 The shattered shell fragments of the earth's crust
Arrows show the main movements taking place at present. Only six major plates are needed to account for all of the earth's major landforms.

Figure 4.2 The Andes mountains

The Andes form where a slab of ocean crust dives under a continent. Sediment deposited on the ocean floor is dragged along and scuffed up against the leading edge of the continent. The ocean crust slowly scrapes against the continent as it sinks, and earthquakes result. Eventually melting occurs and some material rises up as "bubbles" of magma to produce volcanoes and batholiths. The western mountains of the USA and Canada have formed in the same way as the Andes.

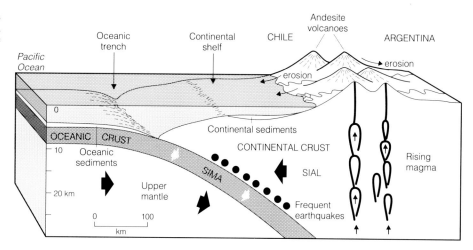

Oceanic trenches

Way down below

One of the most important results of plate movement occurs when one plate collides with one of its neighbours. When this happens there will be either a massive crumbling of the edges of the plate or one plate must ride over the other. There seems to be evidence for both of these types of collision. Indeed the results of overriding were first discovered by a British ship.

Towards the end of the 19th century, the HMS *Egeria* was completing the task of surveying the ocean floor around the island of Tonga in the Pacific Ocean. The crew were amazed that over 8000 m of wire was payed out before the weighted line reached the ocean floor. They had discovered a part of the ocean that extends farther below sea level than Mount Everest rises above it! In fact, further exploration revealed that the Tonga Islands hide a yawning chasm nearly 10 km deep in places.

We know that this gash in the ocean floor — the Tonga-Kermadec trench — is so deep that seven Grand Canyons could be piled on top of each other in it; its length would extend from San Francisco to the Arctic Circle (a massive 3000 km) and yet it is sometimes only 8 km wide.

The Tonga-Kermadec trench is just one of a whole series of trenches that are to be found along the margins of many oceans (Fig. 4.3). They are all near to lines of volcanoes and earthquakes and so are related to active movements of the earth's crust. In fact, trenches are thought to be the surface evidence of the places where a rigid ocean plate is overridden by a rigid continental plate (Fig. 4.2). As the ocean crust sinks down beneath the continent it is remelted along the line of the trenches and part of the molten material rises back to the surface to form volcanoes. Earthquakes occur as the top of the sinking ocean plate grinds against the bottom of the continental plate.

Here we have the means of explaining the origin of at least some mountains. For example, not only has the collision of part of the Pacific Ocean crust caused the deep Chile trench to form, but in the process the edge of the South American continent and the coastal sediments have been crumpled and lifted up into mountains. The sinking ocean plate has partly melted and the magma produced has slowly risen through the continental rocks, over millions of years, as giant bubbles of liquid rock. Today, these rocks have solidified and have been eroded into the jagged peaks of the Andes Mountains.

Exercise 4.1   Ocean trenches and ocean islands

1. Copy Figure 4.3, which shows the deep trenches surrounding the Pacific Ocean. Most of the trenches are named after the islands which they parallel. Using your atlas, name the trenches 1-5.

2. Find, name, and mark onto your copy several large mountains situated within these island chains. You can expect all the mountains you name to be active volcanoes.

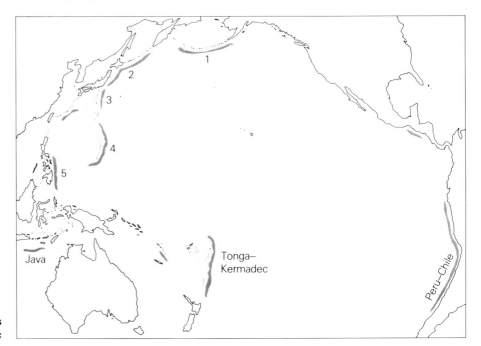

Figure 4.3 Ocean trenches of the Pacific

Fold mountains

Crashing plates

Not all mountains form at ocean/continent margins. The Alps and Himalayas are well away from oceans and so must have been formed in some other way. In these cases scientists believe that two continents have collided. Because both continents are very thick it is not possible for one to be pushed below the other. As a result, when continental plates collide, they mostly crumple up along their leading edges (Fig. 4.4) welding two continents together into a larger unit. India is now known to have collided with the southern edge of Asia, driving up sediments into the great ranges of the Himalayas and the Kunlun Shan (Fig. 4.7). Similarly, Africa appears to have collided with Europe and so crumpled the Alps and the Atlas mountains.

It is important to realize that these are not just events of the past. They are happening now. India is still thrusting into Asia at the same rate that it has done in the past and there is no reason to suppose that the Himalayas have finished building. The great volcanoes of Italy tell us that the Alps are still

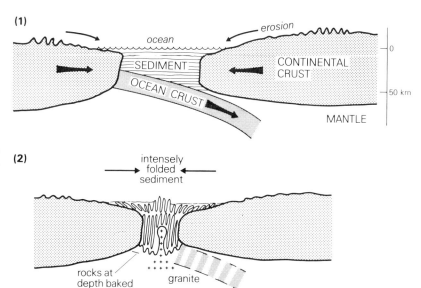

Figure 4.4 How mountains like the Alps and Himalayas were formed
(1) Erosion of continents produces much sediment which is trapped as the continents collide. (2) Collision crushes sediments and some become metamorphosed. Granites also form deep in the mountain "roots."

active and that the Mediterranean Sea may be only a temporary feature. It may yet be lifted up to the height of the Tibetan Plateau which itself was a sea only a few million years ago.

Exercise 4.2 Comparing past and present volcanic landscapes

Scientists can measure the way Iceland is growing as the two halves pull apart and lava wells up to fill the gap in the crust.
1. List the features of volcanicity of Iceland (Fig. 4.5b).
2. List the features of volcanicity of Newfoundland (Fig. 4.5a).
3. Now list the ancient volcanic features of northwestern Britain and Ireland.
4. Write a couple of sentences to say what might have happened to the British Isles and Newfoundland to cause the volcanic activity.

Figure 4.5 Some similarities between Newfoundland, Iceland, and Britain

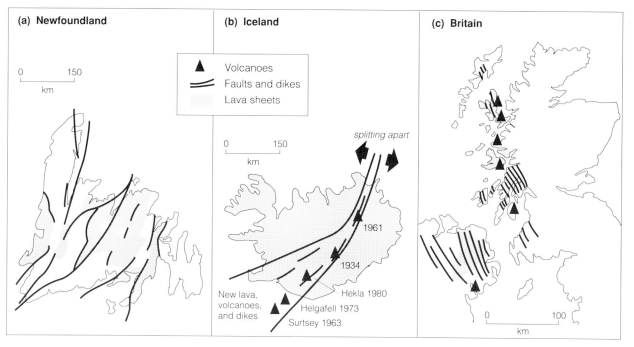

Topic   Metamorphic rocks

Metamorphic rocks are those that have been changed from their original state by the action of high pressures and/or temperatures. For example, the rise of hot molten magma associated with volcanoes and batholiths can bake the surrounding rocks into new rocks with new minerals. However, these only create small baked zones a few kilometres wide at most.

The major regions of metamorphic rocks are found where there have been plate collisions. When continents collide the ocean sediments between them are crushed. As a result the sediments first begin to fold and then to melt over regions that may be thousands of square kilometres in area. The main fold mountains are mostly made from metamorphic rocks that were once sediments in ocean basins.

The type of metamorphic rock depends on the degree of heat and pressure. Sometimes the mass of overlying rock alone can cause rocks to change. However, some of our most attractive banded rocks (called schists and gneisses) and some of our most valuable and attractive minerals (e.g. garnet, ruby) form under conditions of intense heat and pressure. Metamorphic rocks are nearly always very hard and resistant to erosion.

Mid-ocean ridges   ## Welding the cracks

Because some plates are colliding into others, the trailing edge of a moving plate must leave a huge "wound" in the crust as it moves away from its original position. This wound must be healed continually by the rise of new magma along fissures in the crust. The volcanic ridge through the centre of the Atlantic Ocean is one place where healing is taking place today. Because North America is slowly drifting away from Europe, there is a continual need for new lava to well up and fill in the gap in the ocean floor. We see this as outflows of lava along fissures such as the one that caused the Helgafell eruption (p. 9). Thus, just as when continents move towards one another oceans shrink, at the same time new oceans are opening in the places from which the continents came. And because magma heals the fissures, all the ocean floors are made of igneous rock. They are covered by thick layers of sediment only near the edges.

Continents of the past   ## Slow plate to China

We can now see the whole of the earth's complex development in one picture. Volcanoes and earthquakes are mostly the signs of broken plate margins, and mountains and ocean trenches show where plates have collided. Looking at the distribution of young and old fold mountains (Fig. 4.1) enables us to build up a picture of the shapes of past continents. It seems that sometimes plates were thrust into one another only to be pulled apart again and sent drifting off in new directions when convection currents changed. As a result new mountains have formed constantly through time. And as new mountains were formed so old ones were eroded and the sediment taken out to sea to form new layers of sedimentary rock. It is a long story and it is still continuing. The sediment now building up in the Gulf of Mexico will probably be crushed up into mountains of the future, and there are already signs that Africa is splitting away from Asia — the Red Sea is a new ocean being born.

SUMMARY: HOW MOUNTAINS AND OCEAN BASINS ARE FORMED

Most of the earth's surface is covered by six large slabs of crustal rock called **plates**. Each plate may contain a piece of ocean floor as well as a piece of continent. All plates are moved by the drag of convection currents in the earth's interior. As each plate moves, so it must collide with at least one of its neighbours, causing mountains to form. Ocean basins form where two continents split apart.

(a) If two plates carrying continental crust meet, they crush up any rock between them into new fold mountains (e.g. Himalayas, Alps; Fig. 4.4). Any ocean crust is forced deep back into the earth to be remelted. Some of this melted rock rises to the surface again to create volcanoes (e.g. Etna and Vesuvius in Italy).

(b) If a plate carrying a piece of continent collides with a plate carrying only ocean crust, the thin ocean crust is usually forced below the continent leaving an ocean trench and a range of largely volcanic mountains (e.g. the Andes, Fig. 4.2).

(c) Where two plates split apart they leave a wound that is filled with igneous rock giving a thin crust (Fig. 4.6). Water flows in over this new thin crust to form a new ocean. The Red Sea is an example of an early split; the Atlantic Ocean is a split 140 million years old. Very often the centre of this ocean will be the site for a string of new volcanoes (e.g. Iceland, Fig. 4.5b).

Figure 4.6 How the Atlantic Ocean formed
(i) Convection currents develop below a large continent consisting of America, Europe, and Asia. (ii) The continent is split and magma flows to the surface creating a mid-ocean ridge of volcanoes like Iceland. The continents then drift apart.

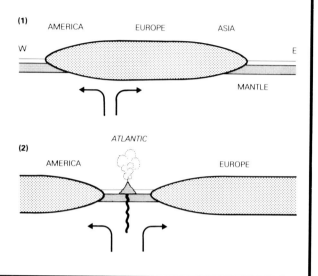

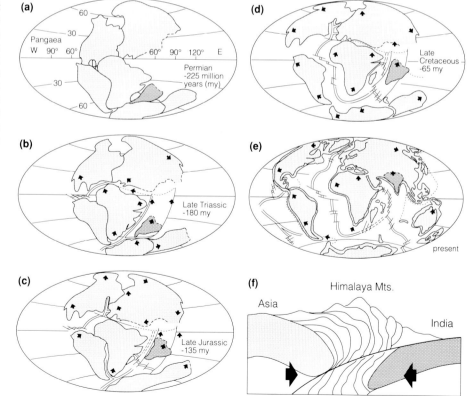

Figure 4.7 Continental drift
The break up of Pangaea is shown in five stages. India moved about 7500 km at rates of as much as 15 cm/a but averaged about 4 cm/a. India's collision with southern Asia caused the upheaval of the Himalayas. The present composition of the Himalayan region is summarized in Fig. 4.7f.

Geology and scenery in Canada

Canada's battle scars

We cannot even begin to understand the present landscape of the world without some knowledge of past plate movements. Without that information, the arrangement of different rocks, mountains, plains, shields, and oceans appears random and meaningless.

Canada occupies the northern half of the complex, ancient North American Plate which is still migrating westward at a rate of 2 to 4 cm a year. The plate consists of some of the earth's most ancient rocks, and huge deposits of some of our youngest rocks. Only 150 million years ago, the plate separated from the super-continent, Pangaea, yet our land bears clear scars of several previous collisions. For example, the Appalachians grew into great mountains about 375 million years ago, probably when the North American Plate collided with western Europe. But there are much older mountains in North America dating back 2300 and 1700 million years ago when a chain of peaks developed along an ancient shoreline near Great Slave Lake. Evidence of such a long series of mountain building supports a modern theory that super-continents form, break up, and reform in a long slow cycle.

The mountains of Atlantic Canada and the eastern USA have weathered and eroded to mere shadows of their original grandeur. At their zenith, perhaps 275 million years ago, those mountains must have looked very much as the Himalayas do today. Now the gently rolling hills of Atlantic Canada stand as the granitic stumps of massive mountains.

When the North American Plate separated from Pangaea, little by little the Atlantic Ocean opened up. Basalt welled up to fill the rift created by divergence, and to fuel the "conveyor belt" upon which the Americas rode. The ocean filled the void and sea levels were maintained. Eastern Newfoundland was torn away from Africa to become a new part of the moving continent. The western margins of the North American Plate were occupied by a vast

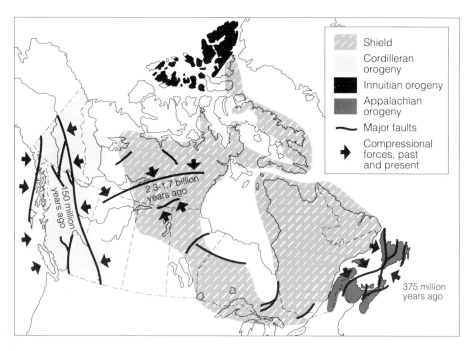

Figure 4.8 Canada Today
The Canada we know today is the result of a long series of geological processes and continental collisions that we can trace back through 4 stages of mountain building and destruction to 2.3 billion years ago.

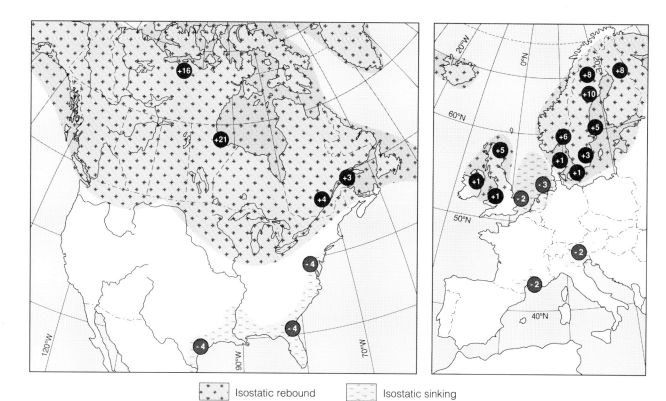

Isostatic rebound   Isostatic sinking

Figure 4.9 Isostatic readjustment of North America and Western Europe
During the past 18 000 years, deglaciation has caused Canada, northern Britain, and Scandinavia to become lighter and float higher. At the same time, deepening oceans and heavy sedimentation caused coastlines near major rivers to sink. The rate of rising (+) and sinking (−) is measured in millimetres per year.

geosynclinal sea, filled to great depths with the sediments of hundreds of millions of years. The **geosyncline** and its sediments were crumpled in the collision between the Continental Plate and the bed of the Pacific Ocean. The subduction of the Pacific Plate beneath the westward grind of the continent caused earthquakes and volcanic eruptions and the compression and metamorphosis of sediments to form the spectacular Monashee, Selkirk, and Purcell Ranges. The Rockies grew from complexly folded and faulted sediments. And, in what appears to have been an east-west, push-pull sequence, the vast rift valley of the Rocky Mountain Trench was formed. The Coast Range grew and eroded quickly so that we now see its core of granitic batholiths in many places between Vancouver and Alaska.

To the east of the Rocky Mountains, the sedimentation of the vast geosyncline, which stretches as far as the Canadian Shield, had been well advanced before the break up of Pangaea. Now another period of deposition has increased the depths of those sediments which form traps for oil, natural gas, coal, and potash to depths up to 7000 m.

The rate of geological activity

A bit of peace and quiet?

Today when we look around, it appears that all the turmoil of events in the past is now over. However, this is far from the truth. India is still colliding with Asia, and Africa with Europe, at just the same speed as they have done now for tens of millions of years. Mountains are still forming even though we can rarely see the changes in our short lifetimes. For example, the whole of Malaysia has been formed in the past hundred million years. At this moment, the southeastern part of USA is sinking by about 4 mm each year, while the Hudson Bay region is rising by up to 21 mm a year. Similarly, northern Britain is rising at the rate of about 4 mm a year while the lowlands of southeast England and the Netherlands are slowly sinking. This process of **isostatic readjustment** will drain all the water out of Hudson Bay in only a few thousand years while the people who live on the subsiding coasts will carry on a constant battle with the relentless sea. Global warming will only quicken the flooding of coastal lowlands.

QUESTIONS

Multiple choice

1. Most oceanic islands have one thing in common. They are:
   (a) surrounded by coral reefs
   (b) sinking as plates migrate
   (c) volcanic in origin
   (d) covered by rain forests
2. What is the most logical cause-effect sequence for tectonic events?
   (a) subduction — volcanic eruption — sea floor spreading — earthquake
   (b) sea floor spreading — subduction — earthquake — volcanic eruption
   (c) volcanic eruption — earthquake — subduction — sea floor spreading
   (d) earthquake — sea floor spreading — subduction — volcanic eruption
3. Select the illogical cause-effect matching:
   (a) rift valleys — crustal compression
   (b) normal faults — crustal tension
   (c) San Andreas fault — tear fault
   (d) synclines — crustal compression
4. Coal, oil, natural gas, and potash are usually to be found in rocks which are:
   (a) volcanic
   (b) sedimentary
   (c) metamorphic
   (d) plutonic
5. Coral atolls form:
   (a) around active volcanic islands
   (b) where sea levels are falling
   (c) where there is no ocean current
   (d) on top of submerged volcanoes
6. Ore bodies are frequently associated with the condensation of minerals near:
   (a) igneous intrusions
   (b) normal faults
   (c) rift valleys
   (d) eroded anticlines
7. Batholiths create distinctive, rugged mountains because they are composed of:
   (a) obsidian and other fine-grained rocks
   (b) homogeneous, coarse-grained plutonics
   (c) jointed and faulted sedimentary rock
   (d) basalts and associated extrusive rocks
8. Canada and Scandinavia are rising while Florida and the Netherlands are sinking. This process is known as:
   (a) inversion of relief
   (b) subduction
   (c) weathering and erosion
   (d) isostatic readjustment
9. The Grand Canyon provides geologists with an ideal laboratory because:
   (a) it has been eroded through 2000 m of lava flows
   (b) its oldest rocks formed more than 3 billion years ago
   (c) 1 billion years of sedimentary rocks are easily accessible
   (d) only geologists are allowed to explore its rock formations
10. Andesitic volcanoes are spectacular and potentially dangerous because they erupt very irregularly and:
    (a) they usually vent through long fissures
    (b) they emit only clouds of cinders and ash
    (c) their lava solidifies very quickly
    (d) their lava is exceptionally fluid
11. Continental collisions and super-continent break up appear to occur in cycles over very long periods of time. The mountains formed by successive collisions can be classified according to their age. A correct order from oldest to youngest would be:
    (a) Himalayas — Appalachians — Andes
    (b) Rockies — Himalayas — Appalachians
    (c) Andes — Rockies — Appalachians
    (d) Appalachians — Rockies — Himalayas

Short Answer

1. Write a 3-page report on one of the following topics: (a) volcanicity, (b) folding, (c) faulting, (d) mountain building. Illustrate your essay with suitable diagrams.
2. Name an area where geysers and mud pools occur. Explain the operation of a geyser, and explain how hot water is used to generate electric power.
3. Name three fold mountain ranges, two volcanoes, and one rift valley. Draw diagrams to show (a) a rift valley and (b) a region of folds showing synclines and anticlines.
4. Explain why some volcanoes erupt violently and have steeply sloping cones, while others have gently sloping cones and produce mostly lava flows. Name examples of each type of volcano.
5. Name a coal field and describe the geological conditions that have affected the mining of the coal. Draw a diagram to show how coal and natural gas are related in the rocks, and explain why both gas and oil are found only in small isolated areas. Describe how this has affected the exploitation of the world's oil and gas reserves.
6. Explain how the following materials are used: limestone, potash, iron ore, and clay. For any two, identify a major area of extraction and describe the way the materials influence the location of industry.

Part two Atmosphere and ocean

The circulation of air and water is a never-ending cycle driven ever onwards by the sun.

Figure 5.0 The swirling pattern of cloud as seen from space

5 The weather machine

The origins of meteorology

The Gunpowder Plot

In the Middle Ages bell ringing was a dangerous job. Bells were rung in church towers and steeples to disperse the thunder — then regarded as the work of the devil. Unfortunately bell ropes kept getting wet during thunderstorms and wet ropes hanging from bells in tall towers provided an easy path for lightning to reach the ground. Bell ringers kept being killed — from electrocution!

Stormy weather was a major hazard to armies and navies as well. There was a constant danger that a flash of lightning might strike a gunpowder store and cause an explosion. During thunderstorms navies very often had to disengage at the height of the battle because of the risks of their ships blowing up from "natural causes."

One day a lightning flash hit a gunpowder store in northern Italy and blew it to bits. It was this event that led to the widespread use of lightning conductors on buildings. At the end of the 18th century it was even possible to buy ladies' hats and gentlemen's umbrellas complete with their own lightning conductors (Fig. 5.1).

But lightning was not the only natural hazard that a ship at sea might have to face. In 1854 the French navy was struck by a hurricane-force wind while anchored off the Crimea and a ship was lost with all hands. This was too much. A means of predicting the weather and especially the possibility of storms was urgently needed.

Figure 5.1 Lightning fashion!

Early weather measurements

All hot air

In the meantime many other people became interested in the atmosphere. Hot air balloons had been flying since 1783. In 1803 two people in Germany rose up to the staggering height of over 7000 m in a balloon, discovering that temperatures decrease with height almost to the point of freezing them. In 1823 the British Meteorological Society was formed to try to organize all the enthusiasm of the new amateur meteorologists and in the 1850s the first weather maps were produced. Now people were able to see how the weather changed over the earth's surface and, to a limited extent, up in the air. The science of **meteorology** had arrived.

It seems a far cry from bell ringers and hot air balloons to the computer-made weather maps we see today. Satellites orbiting above us now provide weather pictures on our television screens within a couple of hours of their being taken. Ideas are changing fast. For example, we now know that ground surface measurements of the atmosphere are not enough. What is happening near the ground is really controlled by events high up in the sky where violent winds flow in north-south waves and drag our weather with them. Weather predictors might still make some errors but forecasts are becoming more and more accurate all the time.

The earth's heat balance

A balancing act

Over the past few years the Mariner and Voyager spacecraft, sent out on exploration missions from earth, have greatly increased our knowledge of the planets. Nevertheless, even without the pictures that are now being transmitted to us through millions of kilometres of space, we knew that sunbathing on the sun-facing surface of Mercury would be, to say the least, a somewhat uncomfortable affair. Mercury is the planet nearest to the sun and, rather in the same way as when we sit too close to a fire, it gets uncomfortably hot. The surface temperature at lunch time on Mercury is a little over 400°C so there would be no trouble roasting a turkey even without an oven! By contrast, a few astronomical steps will take us to one of the outer planets, Saturn. Here the sun appears as just a bright star in the sky and its heating effect is so small that the "surface" temperature is as low as −150°C.

So distance from the sun determines the energy received by the planets. Fortunately the earth is at an intermediate distance and the solar energy we receive (called **insolation**) neither roasts nor freezes us.

Storing solar energy

Greenhouse earth

In the long term, the earth's surface and atmosphere seem to be in balance, becoming neither hotter nor colder. The surface heats up because it receives energy from the sun. The earth in its turn radiates the received energy back out to space until there is a balance between income and output. But whereas the atmosphere is fairly transparent to incoming solar radiation, a lot of the outgoing radiation is temporarily trapped by water vapour and carbon dioxide gas in the air. As a result the atmosphere receives an extra source of heating (Fig. 5.2).

The role of water vapour is seen very well by comparing clear and cloudy nights. On cloudy nights the cloud "blanket" absorbs heat from the earth and radiates it back again; as a result the weather is much less cold. Even when clouds are not present, water vapour and carbon dioxide in the air still prevent some of the heat loss. Indeed, the way the atmosphere allows solar radiation in, but partly stops it from being lost again to space, is superficially so like the way glass helps heat a greenhouse that it is called **greenhouse effect.** In fact the greenhouse effect is responsible for keeping air temperatures 25°C higher than they would be otherwise.

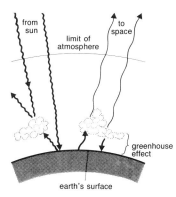

Figure 5.2 Heat balance of the atmosphere Energy coming from the sun balances energy lost to space over a year. However, temporary absorption of heat by clouds and air keep the atmosphere warm.

Changes in the sun's heating power

Rays of warmth

There is a lot of energy available in the rays from the sun. Sometimes this is dramatically illustrated, no more clearly so than in the hills of southwest France, near the town of Odeillo. Here, high up in the Pyrenees, a hillside has been covered with stainless steel sheets adjusted in such a way as to make a vast concave mirror. These mirrors are slowly turned and the sun's rays, caught by each shimmering plate and focussed on a single point, instantly provide enough heat to melt steel (Fig. 5.3a).

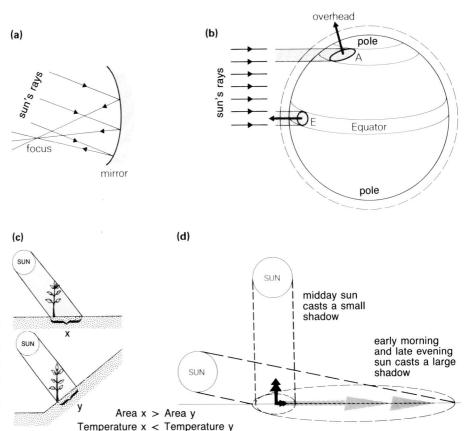

Figure 5.3 Examples of energy concentration
(a) Concave mirrors concentrate the sun's rays. (b) At E, near the Equator, the sun is overhead and energy is not spread out; at A, nearer the pole, the sun is not overhead and the same energy that heats E now has to heat a greater surface area. As a result A is warmed less than E. (c) When land surfaces are tilted, the same sun's energy is used to warm a smaller area of ground and so crops grow more readily. (d) Long shadows in the morning and evening show how the sun's energy is being shared over large areas. As a result the sun's rays appear to be less warm than at midday. The same reasoning applies to a comparison between summer and winter.

On the earth's surface the most concentrated rays occur when the sun is overhead and they become progressively more spread out as the sun's angle to the vertical increases (Fig. 5.3b). It is this that makes the earth much hotter at the Equator than at the pole, and midday hotter than early morning. On a local scale, the amount of heat received depends on the direction and tilt of a hill slope. Farmers in upland areas all know the importance of slope and aspect to their activities (see Ex. 5.1).

Exercise 5.1 The importance of aspect

Aspect is the compass direction that a slope *faces*. For example, a slope that faces due south has an aspect of 180°, southeast slopes 135°, etc. A slope's inclination is also important. The noon sun shines most directly upon slopes that have the same angle as the latitude of that slope. For example, at an equinox at noon, a 180° aspect slope with a 50° inclination at 50°N latitude will receive the sun's most direct rays.

1. Copy the map (Fig. 5.4) and mark on the regions covered by vines and forest.
2. Draw a circle with a diameter of 5 cm. Assume the "12 o'clock" position is north. Now use a protractor to find the 4 most northerly facing slopes still used for vine growing. Draw these aspects as radial lines onto your circle. This indicates the limiting conditions for vine growing.
3. Find the direction of the slopes (A, B, C, D, & E) where vineyards are growing farthest from the Rhine. Mark these 5 radii onto your circle.
4. Using the results of your drawing, describe how aspect controls the growth of vines in the gorge.
5. From the diagram and Figure 5.3 suggest another physical factor that might be important in controlling the growth of vines.

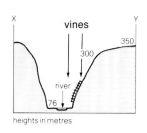
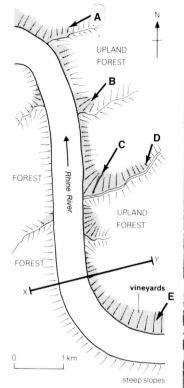

Figure 5.4 The Rhine gorge (Germany)
Right. In plan and, *above,* in section. *Far right.* Vines growing on a terrace.

SUMMARY: HEATING THE ATMOSPHERE

The atmosphere is primarily heated from below. Solar **radiation** warms the ground. Air in contact with the warm ground is warmed by **conduction**. The warmed air is less dense than the air above and it rises through it by **convection**. Winds also bring air across the ground surface, mixing warm and cold air together by **turbulence** (Fig. 5.5). When clouds form more heat is provided to the air by **condensation**.

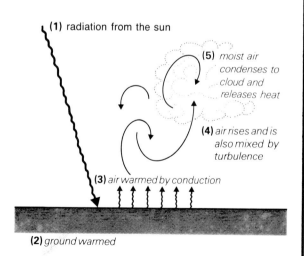

Figure 5.5 The ways the atmosphere is heated

Air circulation patterns

The restless atmosphere

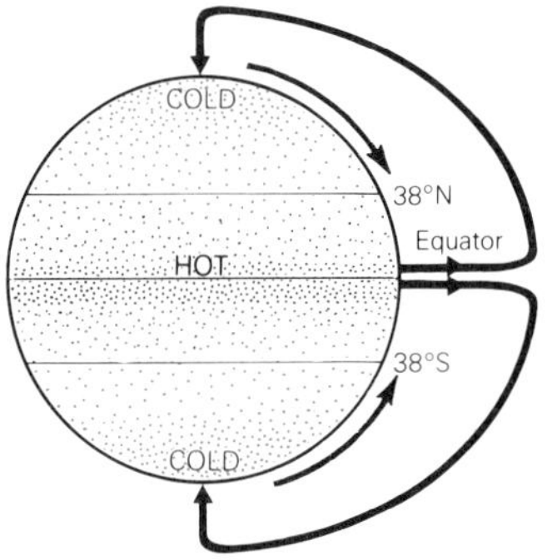

Although a long-term balance is achieved between earth's input and output of energy as a whole, all areas polewards of about 38° of latitude lose more heat to space than they receive from the sun, and all areas equatorwards gain more than they lose (Fig. 5.6). So why don't the poles get colder and colder and the tropics hotter and hotter?

The answer is provided by the atmosphere and oceans, which share out the surplus solar energy of the tropics and moderate the extremes of temperature that we might otherwise expect.

You can get a good idea of how the oceans and atmosphere function because the same processes operate in many homes. In the living rooms of most mid-latitude homes there is an equivalent to the tropics — a heat source such as an open fire, airtight stove, electric radiator, or a forced-air vent (Fig. 5.7). In fact, although they all **radiate** energy out into the room, a large part of the energy is used in warming up the air in contact with them by a process called **conduction.** The warmed "jacket" of air surrounding the heat source then expands to become less dense than the surrounding air and it rises, causing fresh cool air to take its place and be warmed in turn. This is just the same way that air circulates in the atmosphere.

In such living rooms the coldest surface is usually the window. Here the "polar region" of the room, air is colder, more dense than its surroundings and, therefore, it sinks. The heat source and the window together set up a circulating or **convectional** flow of air around the room. The tropics provide the heating source for the earth's atmosphere, and oceans leaving the cold polar regions help to provide the temperature contrast which sets up oceanic convection. As a result, there is a tendency for two basic convectional flows to be established around the earth to share out the heat: one from the Equator to the North Pole and one from the Equator to the South Pole (Fig. 5.6). They are more or less mirror images of one another.

Figure 5.6 Convection in the atmosphere
The earth's surface receives more heat than it loses equatorwards of 38°, and less polewards. If the earth were stationary, a simple convection system of one cell for each hemisphere might develop as shown.

The effects of the earth's rotation

Turn, turn, turn

So far we have not considered how the rotation of the earth might affect convection. You will already realize how difficult it is to cause liquids and gases (fluids) to move in straight lines — even water gurgling out of a bath will always swirl round and round. Fast-flowing rivers are another example: not only do they flow in a turbulent way, but they insist upon following a winding path. Whenever wind whisks up leaves from a pavement they are lifted in curved paths. So we should expect the atmosphere and oceans to behave in the same way.

Sometimes this is easy to see and tornadoes, typhoons, and hurricanes are well known for their spiralling winds (Fig. 5.25). Less well known but much more important are meandering (winding) tunnels of air high up in the sky (called **jet streams**, p. 56) and swiftly moving tunnels (streams) of water set within the oceans called **ocean currents** (such as the famous Gulf Stream, p. 82). Wherever you look, gases and fluids are on the move; we saw this to be true even with the convection currents within the earth (p. 7).

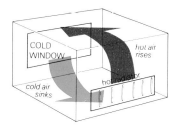

Figure 5.7 A room convection cell
When heating source and window are on opposite walls, cold drafts are generated around the feet. Can you think of better places for the heating source to be to prevent drafts?

Figure 5.8 Local-scale convection along a coast

Exercise 5.2 Local-scale convection

1 Look at Figure 5.7. Explain why many people say they get hot heads but cold feet while sitting in a heated room such as this.

2 Builders place radiators underneath windows. What effect would this have on convection currents? Would it be a good or a bad thing?

Exercise 5.3 Coastal breezes

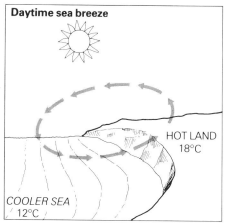
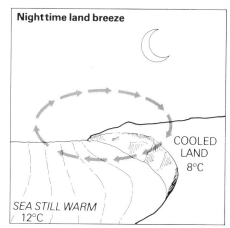

Convection operates at all scales from the size of a room to the size of the earth. Figure 5.8 shows the convection that often occurs along sea coasts.

1 Why in the summer is the daytime breeze called a sea breeze and at night called a land breeze?
2 Sea water absorbs a lot of heat in its upper layers and keeps a fairly constant temperature day and night. By contrast land can store very little heat, so it becomes very hot by day but loses heat rapidly at night. Using the temperature figures on Figure 5.8, explain why convection changes direction between day and night.
3 Would Figure 5.8 still be correct in winter?

These swirling patterns are enormously important in that they control our daily weather. Their effect is easily seen by comparing the simple pattern outlined in Figure 5.6 which was drawn for a "stationary" earth with the picture of the real spinning earth shown in Figure 5.0. The basic convection currents are still the main driving force of the circulation. That has not changed. But on a spinning earth the most effective way of transferring air from one place to another is to use the apparent chaos of giant swirling movements.

High and low pressure

What goes up must come down

If you look at a diagram of the atmosphere (Fig. 5.10) you will find that over some parts of the earth the air is rising, while in others it is sinking. On a global scale, however, all the ups are balanced by the downs, despite these regional differences.

Sometimes the air sinks faster than it can push the air below out of the way. It is rather like a lot of people trying to leave after a hockey game: they all crowd together as they try to push through the gates. In the atmosphere the build up of air is called a **high pressure region.** Conversely air can be drawn away from a region faster than it can be replaced. This gives a **low pressure region.**

There are some parts of the world where air is always building up faster

Topic *Global winds*

Eventually all the air that rises at the Equator and flows at high levels towards the poles has to return near the surface to complete its global circulation.

High in the sky we might expect the air to be streaming polewards. This is shown on the cross section, Figure 5.10. However, if you were to go up to these high altitudes in a plane you would find that the air appears to be moving largely to the east and not polewards at all!

Still air at the Equator is actually moving very quickly. It has to move at nearly 2000 km/h just to keep up with the spin of the earth! You can see this on Figure 5.9a. Any wind we experience is due to small changes from this need to keep up with the spinning earth.

Now suppose the upper air moves from the Equator, where it is going fast, towards the poles, where it only needs to go slowly. As soon as the air leaves the Equator it is going too fast for the latitude it arrives in, so it gains on the earth. So all the time air moves to the pole to share out the sun's heat, it keeps going too fast, turning a simple direct motion into a very long curved path (Fig. 5.9b). In practice, the curve is a very gentle one with air moving faster in an easterly direction than it does polewards. We feel this excess speed as a **wind**. Because we describe winds by the direction from which they come, this is called a *westerly* wind. Conversely, air returning to the Equator is always moving more slowly than the latitude it reaches. We feel this as an *easterly* wind. The surface wind patterns of North America are shown on Figure 5.10.

It turns out that a nice even westerly or easterly wind cannot be maintained on a spinning earth without developing places of slack and places of fast movement. This is just the same as the uneven flow in a river. There are some fast-moving "streams" embedded in the main flow. In the atmosphere some of these have such fast-moving air currents in them that they have been called **jet streams** (Fig. 5.9b). There are two jet streams in each hemisphere. One steers high and low pressure systems over the mid-latitudes. The other is above the region that receives the monsoon. Today weather forecasters keep as close an eye on high-level jet streams as on ground-level observations.

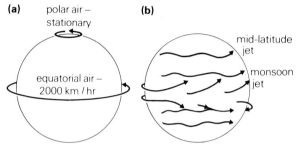

Figure 5.9 Global winds
Air really takes a long curved path on its way to the poles at high altitude.

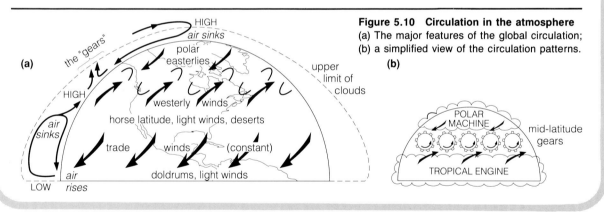

Figure 5.10 Circulation in the atmosphere
(a) The major features of the global circulation;
(b) a simplified view of the circulation patterns.

from above than it can get away at ground level. These are the *permanent* high pressure regions. A series of such regions are found about 30°N and S of the Equator, encircling the earth. Underneath them lie many of the world's great hot deserts. They are the cloudless belts on Figure 5.0. Each polar region is also a place of high pressure, and here we find the great ice deserts.

There is only one region that has permanently low pressure. This is a belt of air near the Equator, which moves seasonally with the overhead noon sun. In this region heavy precipitation is very frequent. Its position is shown on Figure 5.0 as a narrow band of cloud.

You can see these belts again on Figure 5.10 and they show that the atmosphere contains very definite patterns in its air movements, driven, as it were, by the great heat energy of the tropics. In fact the whole global circulation can be thought of as an engine powering a machine. Near the Equator there is the engine, providing the power to move the rest of the system. This power is transmitted through a set of "gears" (the swirling patterns on Figure 5.10) to drive the machine (the polar regions) which continually need energy pumped into them. People in southern Canada live in the region of the "gears" and thus have remarkably variable weather.

When high is low

Changes in air pressure with height

So far we have looked at the way in which convection and a spinning earth have given rise to places of high and low pressure. These determine whether we shall have precipitation or drought. Our next task is to examine the properties of air pressure to see how it is related to precipitation.

Participants in the 1968 Olympic Games at Mexico City knew they were going to have problems. Mexico City is over 2000 m above sea level. The athletes knew they would have to arrive at the Olympic Village several weeks before the games started just to acclimatize themselves to the thin air. Otherwise they would have been gasping for air and would have given poor performances.

Mountaineers also know this problem well: they stay for days at one high-level camp before climbing farther and establishing a new camp. Again it is a matter of acclimatization. Air becomes "thinner" (less dense) the higher you go above sea level. It doesn't change much in composition, but there are just fewer molecules of it in every cubic metre as compared with sea level. Go up in a plane and you will be sitting in a pressurized cabin, because at 10 000 m you cannot breathe in enough oxygen from the thin air to keep you alive. And it is colder too, partly because you are out of the greenhouse region of the lower atmosphere. This is why planes are also heated.

A reservoir in the sky

Forms of water

Air pressure and temperature are closely related and in the atmosphere both usually decrease with altitude. In the first part of this chapter we read that two German balloonists made the first ascent over 7000 m in 1803. Not only did they get very cold, but one of them found that he could no longer get his hat on — his head had become swollen because there was less pressure on it!

Water vapour is one of the gases in the air that is important in the greenhouse effect. It is also important because it can be turned into liquid water droplets. Water droplets are easily seen whereas water vapour is invisible. A mass of water droplets suspended in the air is called a **cloud**. But what causes water vapour to turn into droplets and what has it to do with balloon ascents?

One way of producing water droplets is to cool the air because *the cooler air becomes the less water vapour it can hold.* So if the air temperature is lowered sufficiently the air becomes **saturated** (filled to capacity) with water vapour. Any further decrease in temperature will then result in water being condensed from the vapour state and converted into liquid droplets.

Air cooled to saturation is quite a common phenomenon. This is what happens to the pocket of air around a glass of lemonade when it makes the glass wet on the outside. **Dew** on the grass forms because of the same effect when vegetation cools overnight. But when there is no cold surface for the air to condense on and a whole thickness of air has cooled down to its saturation

Figure 5.11 Cloud formation
A temperature inversion is apparent in the Nechako Valley, BC. Water vapour in the warm upper air condenses to form a narrow band of cloud where it mixes with the unusually cool air in the valley bottom.

the droplets remain suspended in the air as cloud. You produce a "cloud" when you breathe out on a cold day. The warm moist air from your lungs is quickly cooled until water droplets form. Steam from a boiling kettle and clouds being formed above a power station show the same effect.

Topic How raindrops and snowflakes form

In the atmosphere as air rises it cools until it reaches a temperature at which it can no longer keep all of its water content as vapour, and it has to shed some as water droplets or snowflakes. We see the myriad of droplets or snowflakes as a cloud. But each droplet or crystal is tiny — a matter of only millionths of a metre across — so how do they form into raindrops and snowflakes big enough to fall to the ground?

It may come as a surprise to find that most precipitation over non-tropical regions starts as snow, even on a summer's day. Whether you receive snow, sleet (semi-melted ice), or rain largely depends on the air temperature below the cloud. If the air is warm, the falling snowflakes melt to become rainfall at ground level.

Temperatures in a cloud are very low because clouds are so high (Fig. 5.12). In many cases parts of clouds have temperatures as low as $-30°$ to $-40°C$. Because of this, much water vapour is shed directly into tiny snowflakes (see Fig. 17.4 for a typical shape), rather than liquid droplets. As a result the top of a cloud is sometimes all snowflakes while the (warmer) bottom is all water droplets. This leaves a mixture of the two in the middle.

In a cloud, air is continually moving upwards. The upward movement brings with it water droplets. However, in these middle regions temperatures are really too low for the water droplets wafted up from below and so they tend to "condense" on the snowflakes already there, causing the snowflakes to grow. Eventually some snowflakes grow to a size when they can no longer remain suspended by the upward air movement and they fall out of the cloud. So although there are many water droplets at low levels, they often do not grow; it is only the snowflakes in the mixture region that grow big enough to fall out. A deep cloud, with air currents rising to regions high enough to have snowflakes growing as well as a good supply of vapour to produce droplets lower down, looks dark and grey from the ground because its thickness reflects so much light from the sun. This is why we say that dark clouds threaten precipitation but that white clouds are a sign of fair (or at least rainless) weather.

In the tropics, even deep clouds do not cool enough for snowflakes to grow. In these regions, upward air movement is often very fast and water droplets are dragged together. When sufficient droplets have bumped together, very large droplets form. Droplets have to be very large to fall out of a cloud with rapidly rising air currents. This is why many tropical storms have torrential rain falling as large drops.

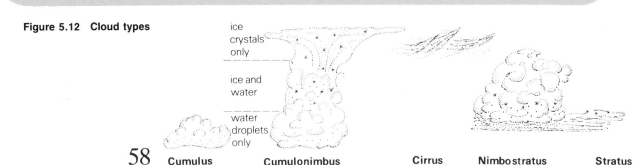

Figure 5.12 Cloud types

Cumulus   Cumulonimbus   Cirrus   Nimbostratus   Stratus

Top. Small **cumulus** clouds only contain water droplets and will not produce precipitation. *Centre left.* Deep **cumulonimbus** cloud has a mixing region of ice crystals and water droplets and will produce heavy rain, thunder storms, and lightning. *Centre right.* High-level **cirrus** cloud contains only widely scattered ice crystals and will not produce precipitation. *Bottom.* This **nimbostratus** cloud is thick enough for ice crystals to form; it produces widespread precipitation typical of a warm front; usually only the lower part of this cloud can be seen from the ground. **Stratus** cloud is thin and it will only produce drizzle.

Cloud formation **Up, up, and away**

When air moves towards a mountain range it has no choice but to go over it. This makes it cooler and brings it ever closer to saturation. At some height up the mountainside saturation is reached and then water vapour is turned into droplets and a cloud is formed (Fig. 5.13). Cloud formed in this way is called **relief cloud,** and any precipitation that may fall from the cloud is called **relief precipitation.**

However, when the air has passed over the highest peak it starts to sink down the other side and, therefore, it becomes warmer. The warmer the air, the more moisture it can hold, and so droplets are no longer produced. We see this as the end of a cloud on the leeward (sheltered) side of a mountain. Sometimes the leeward end of a cloud bank is dramatically clear, like a wall. Also with much moisture lost to precipitation over the mountains there is little left to give any precipitation to lowlands in the lee of the mountains. Low precipitation regions leewards of a mountain barrier are said to be **rainshadow** regions. Where mountains are very high they cause the leeward side to become a desert.

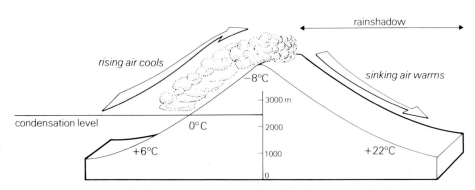

Figure 5.13 Relief cloud
Relief cloud is produced when air is forced to rise over a mountain range. A rainshadow zone extends out on the lee side.

Exercise 5.4 Rainshadows

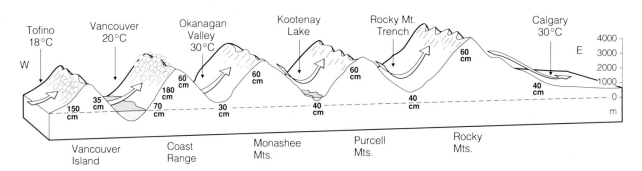

Figure 5.14 Cloud and rain over southwestern Canada

Calgary, Alberta lies to the east of several mountain ranges in southwestern Canada. Here, at latitude 50°N, moisture-laden winds blow predominantly from the west. In your notebook, complete the following passage (in 500 words or so) in order to explain the frequency of chinooks and the cause of a rain shadow at Calgary.

As moisture-laden winds blow in from the Pacific Ocean, they are forced to rise over the rugged Insular Mountains of Vancouver Island. There, orographic uplift causes temperatures to drop rapidly, clouds form, and heavy precipitation occurs. As the air sinks on the lee side, the air warms and the Gulf of Georgia region receives only about 350 mm of precipitation. The air then rises over Vancouver and the Coast Range...

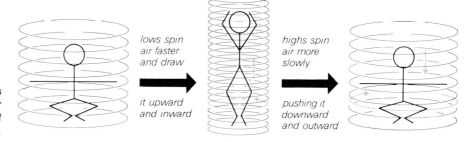

Figure 5.15 Highs and lows
The spinning figure skater illustrates some features of highs and lows.

Frontal clouds

Battle of the masses

Air is forced to rise, and so to produce cloud and precipitation in many circumstances other than over mountain barriers. Most commonly, air rises within the spiralling air stream of the mid-latitude "gears." These spirals are very large — often well over 1000 km in diameter — and are really columns of revolving air stretching up from the ground surface to great heights (Fig. 5.15). The columns not only revolve in different directions (clockwise or anticlockwise) but can either "tighten up" to stretch further upwards or "open out" to become shorter so that the air in them moves more slowly. In many ways this can be likened to ballet dancers and figure skaters standing up and bringing their arms in to increase their spinning, or opening their arms to slow down. Columns of air with downward spirals push air downwards and warm it, which results in a high-pressure region on the ground with clear skies above; we call them **anticyclones**. But those with upward-spiralling air (called **depressions**) pull air up from the ground, creating a low-pressure region and at the same time cooling the upward-moving air to give cloud and precipitation. Depressions are the major source of precipitation in the mid-latitudes.

Because these air columns move over mid-latitudes from west to east, areas such as southern Canada tend to have wet spells when an upward-spiralling

---

Topic Cloud types and precipitation

A cloud is the visible form of millions of minute particles of water or snow in the free air.

Clouds are classified into three main groups (Fig. 5.12):

Cirrus clouds have a fibrous or feathery appearance and usually form as detached white bands. No precipitation falls from these clouds.

Cumulus clouds occur as detached clouds, being dense with sharp outlines. They develop vertically in the form of rising mounds, domes, or towers, of which the bulging upper part often resembles a cauliflower. The sunlit parts of these clouds are mostly brilliant white; their bases are relatively dark grey and nearly horizontal. The rain that falls from well developed cumulus clouds is heavy and it occurs as downpours.

Stratus cloud is generally grey cloud, lying in a widespread level sheet. It often gives prolonged rain, drizzle, or snow.

The word nimbus is added to cumulus and stratus clouds when they become thick, grey, and capable of producing precipitation.

Cumulonimbus clouds produce showers; **Nimbostratus** clouds produce prolonged precipitation.

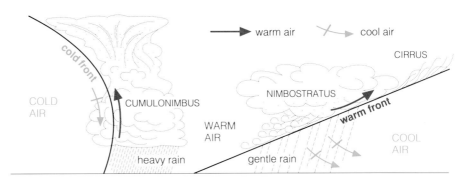

Figure 5.16 Frontal cloud Warm air drawn into a depression rises up over the cold. As it rises it cools and produces widespread cloud banks over the frontal contact region.

depression drifts over, separated by dry spells when a downward-spiralling high-pressure anticyclone brings clear skies. Actually depressions are more complex than simple columns of rain-laden cloud. They draw in air from all directions; they are, after all, the chief means of transferring energy from the Equator to the poles. Air pulled in from the tropics is warm and it contrasts with cold air brought equatorwards from polar regions. It takes time for these very different types of air to mix and so for a long time there are sharp changes across the boundaries (called **fronts**) where they meet.

Because warm air drawn from the tropics is less dense than the cold polar air that it meets within a depression, the warm air will rise faster than the cold as they are both pulled up into the tightening spiral of the depression (Fig. 5.16). As a result its place on the ground is taken by the cold air. Eventually the warm air ends up on top of the cold air, but until this lift is complete the warm air occupies a wedge-shaped area on the ground. The frontal regions show that the biggest lift occurs either side of the wedge of warm air and they are associated with thick cloud belts and precipitation. This precipitation is called **frontal** precipitation and it is mostly from cloud of a stratus type, occurring in belts perhaps 500 km long. Therefore, frontal precipitation affects large parts of regions such as western Canada at the same time. Notice also that this type of precipitation is not dependent on relief, so it falls on lowlands and uplands alike.

No man's land

Cumulus clouds

We have been discussing the way in which moving columns of air are associated with special types of weather: highs with clear weather and lows with fronts, cloud and precipitation. However, between low pressure regions and anticyclones there are quite large areas which are neither one thing nor the other — a sort of "no man's land." There are no fronts in these regions to pull warm air up to make clouds. Similarly the air is not being pushed down, keeping the skies clear. Without any special control from the "gears," the weather is able to develop in a variety of ways — usually fairly unpredictable — and to give a slice more variety to our weather.

Very often this sort of air region has patchy cloud in it and sometimes this cloud can cause precipitation to develop, for this is where we most commonly find cumulus cloud, especially if the air is flowing from the north. Unlike stratus cloud, which results from a widespread organized uplift of air over mountains or at a front, cumulus cloud is the result of local ground and air conditions, each bubble of cloud developing spontaneously at some moment when

Topic Pressure and weather maps

Atmospheric pressure — the weight of air above a particular level — is measured with a **barometer.** Scientists have decided that the reference value by which to judge all pressure values should be called 1 bar. This is divided into a thousand parts, each called a **millibar (mb).** The millibar is the unit used to show pressure on the ground surface charts we call weather maps.

With this system, the lines of equal pressure (called **isobars**) are rather like contour lines on a map of the land surface. Therefore, you can think of a high pressure region in the same way as a hill on a contour map — the highest pressure is in the centre and values get smaller outwards. Low pressure regions are the opposite — more like basins — with the lowest value in the centre.

Air flows from regions of high pressure to regions of low pressure, just like water would run down a hill into a basin. The wind speed depends on the pressure gradient, so the closer the isobars, the stronger the wind. The flow of air is called a wind but the wind direction is not at right angles to the isobars as you might at first suppose. This is because the spinning earth causes the air in both high and low pressure regions to spiral as well, which makes an air stream take a much longer "corkscrew" route from high to low, rather than the steepest and shortest path. Any wind direction on a weather map is therefore at an angle to the isobars, inclined *inwards* to the centre of a low pressure region and pointing *outwards* from a high pressure region (Fig. 5.17).

Weather maps mainly show isobars so that you can work out the way airstreams are flowing. However, they also show temperatures, the type and amount of cloud, and the present weather. This information is conveniently grouped around the wind speed and direction arrow (Fig. 5.18). Very often curved lines are also drawn on these maps to separate regions of quite different types of air. These lines are called fronts and are marked with the special symbols shown on Figure 5.17a.

A depression and part of an anticyclone are shown in Figure 5.19. Here the inward-spiralling air of the depression is picked out by the spiralling cloud pattern. Notice that the warm sector contains much less cloud than at the fronts. Most of the cloud is related to the occluded front. Ahead and behind the fronts are clearer skies with patterns of isolated clouds. Another depression can be seen over Greenland (top left), and France (lower right) is under a high pressure system. The eye of the storm is midway between Scotland and Iceland, the cold front is over Ireland, and the warm front lies across England.

Figure 5.17 Weather systems
(a) Depressions are characterized by inward moving air, the meeting of warm and cold air and the production of fronts, stratus type cloud and continuous precipitation with quite strong winds. (b) Anticyclones are characterized by outward moving air, clear skies and light winds.

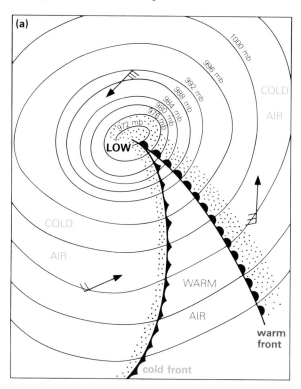
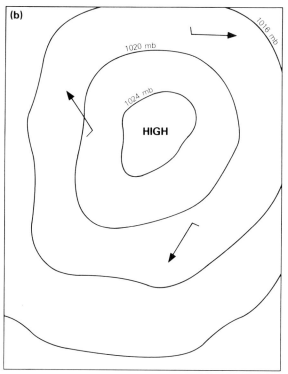

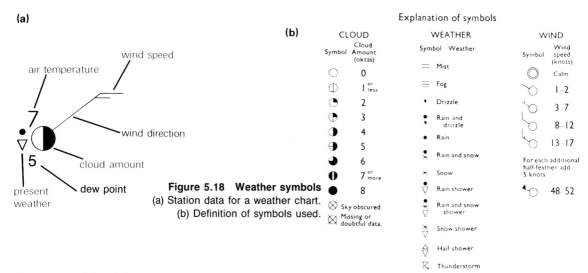

Figure 5.18 Weather symbols
(a) Station data for a weather chart.
(b) Definition of symbols used.

Figure 5.19 A North Atlantic depression seen from space

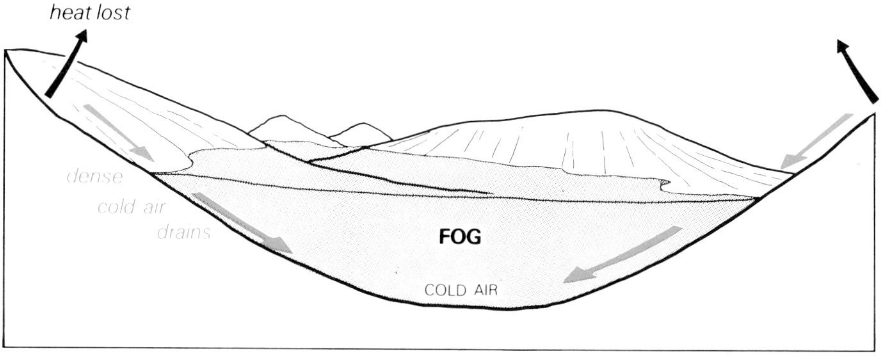

**Figure 5.20
Radiation fog**
This forms on a clear cloudless night in valleys. Hillsides lose heat by radiation; air in contact with cool land cools and sinks into the valley. When temperatures fall far enough for condensation to occur in the air, fog forms. Radiation fog is common under anticyclonic conditions.

Topic Fog

Fog is a cloud on the ground, made from tiny droplets of water. How far you can see in fog depends on the concentration of these droplets: the more droplets, the denser the fog. This in turn depends on the amount of moisture originally held in the air as vapour. A fog is present if visibility is less than 1 km.

Because air over water holds the most moisture, the denser fogs are usually those at sea (advection fogs, see below). However they can be dense inland (radiation fog) especially if added to by smoke and other pollution.

Radiation fog (Fig. 5.20). 5 December 1952 was a very calm cold day over London, and the Thames Valley was full of cold air. As a result a **radiation fog** developed, but because there was no wind and the sun was too weak to warm the air, there was nothing to disperse the fog. It brushed against pavements and buildings and made them damp while all the time factories, houses, and vehicles poured tonnes of smoke and fumes into the air — air which was still. By the next day the pollution accumulating in the cold fog had turned it from grey to brown, and the following day the visibility was cut to centimetres.

Eventually a wind did build up and disperse the pollution-laden fog (called **smog**) but the damage had already been done and at least 4000 people died that week from chest and breathing problems. Here is an example of a natural phenomenon whose effects had been turned into a disaster by people's activities. It was this event that led directly to the creation of clean air zones in British cities where only smokeless fuels may be burned. In Britain fog still occurs, but its effects are no longer so damaging to health.

Advection fog (Fig. 5.21a and b). There is another type of fog called **advection fog** which plagues coastal regions such as California and Newfoundland. It even affects British Columbia's coast. In these cases warm moist air is pulled across cold sea, lowering the temperature of the air until droplets of water are released to produce fog. If the air blows from the sea onto the land, inland the sun will warm the air again, dispersing the fog. If warm air moves from land to sea, the fog bank forms some kilometres off the coast and is not so readily dispersed.

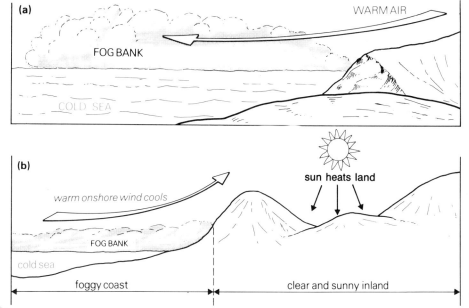

**Figure 5.21
Advection fog**
This forms (a) over seas or (b) along coasts when warm moist air travels across a cold surface. This type of fog is not easily dispersed by winds because they just bring more warm air to be cooled by the huge store of cold water. The famous fog banks off the coast of Labrador can last for months.

65

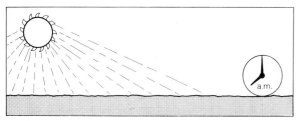

(1) Clear sky; sun begins to heat ground. Some parts become hotter than others.

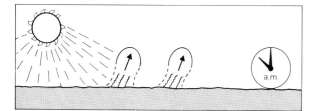

(2) Bubbles of air in contact with hotter ground start to rise.

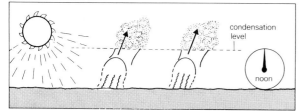

(3) Rising air bubbles reach the condensation level and cloud forms. Bubbles continue to rise, making clouds deeper; new bubbles become heated.

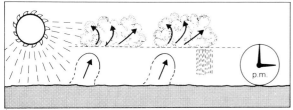

(4) More and more clouds form and begin to obscure the sun. Some are deep enough to give showers and perhaps thunder.

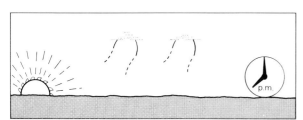

(5) Sun sets, ground cools and the last of the heated bubbles "melt" away into the surrounding air.

Figure 5.22 The formation of cumulus clouds

conditions are right (Fig. 5.22). The "right" conditions are numerous, but we will consider one of the causes of cumulus cloud that occurs on some fine and sunny summer mornings when the air is crisp and the visibility "too good to be true." On such mornings the first cumulus clouds usually appear by about 10 a.m. and gradually enlarge through the morning until by mid-afternoon the sky can be virtually obscured. As late afternoon approaches, the possibility of showers or even thunderstorms looms large, but with the onset of evening the clouds seem to evaporate, get smaller, and eventually fade away to give a clear sunset. Many of these clouds are in some way associated with the daily surface heating effects of the sun because they get bigger as the sun rises in the sky and disappear as the sun sets and the heating is lost. Here, the convection system is at work again, this time causing only local bubbles of air to rise as they are warmed. Cumulus cloud produces showers. The precipitation is, not surprisingly, called **convectional precipitation**. It is this type of precipitation, together with the effects of frontal precipitation and relief, that determines the total precipitation at any place in the mid-latitudes. However, the really important regions for convectional precipitation are nearer the Equator where frontal effects are missing and surface heating more intense.

Topic *Thunder and lightning*

Cars and trucks must generate over 8000 volts just to create a spark a few tenths of a millimetre long and explode gas in a cylinder. Just imagine how much energy and how many million volts must be involved when a flash of lightning sends a spark thousands of metres through the air and causes great reverberating rumbles of thunder (Fig. 5.23).

When air carries a heavy current (lightning), it gets hot just like an electric wire. The air very near a lightning flash heats to about 30 000°C and, as a result, expands as a shock wave. When the transmitted energy of this shock wave reaches our ears, we hear it as a clap of thunder.

To produce lightning a negative charge of about 300 million volts must build up in the cloud. We think this begins in the lowest levels of deep cumulus clouds where snowflakes form. The earth becomes positively charged below a thunder cloud, so eventually a spark jumps the gap between cloud and earth in a series of jerky steps. In fact a lightning flash is many of these jerky flashes separated by only milliseconds of time, so we think we see only one spark, but each jerk and flash generates a shock wave and we hear these as a confused "roll" of thunder. Also because sound waves travel slowly and the flash might be 5 km long, sound from the heated air near the cloud will take longer to reach us than the shock wave from the same heated air near the ground. This, too, makes the thunder last longer.

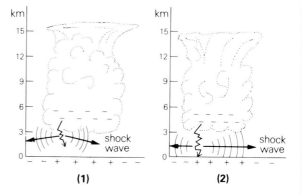

Figure 5.23 How lightning and thunder are produced
(1) A spark jerks down toward the ground, heating air and sending out shock waves as it descends. (2) The spark reaches the ground and the flash is completed. Heated air near the ground sends out further shock waves. These are heard as thunder.

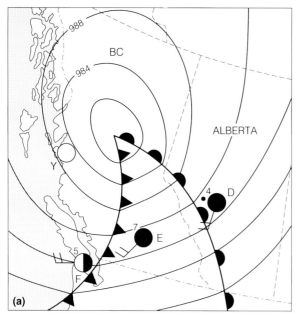
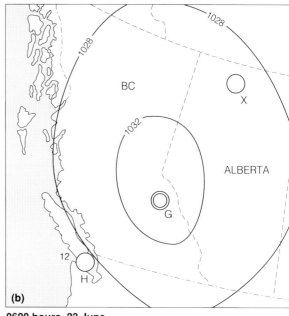

(a) 0600 hours, 13 January

(b) 0600 hours, 23 June

Figure 5.24 Weather charts for January and June

Exercise 5.5 Weather charts

Study the two weather charts in Figure 5.24. Make a careful note of the time of year which each chart shows. Think about the consequences of this before you answer the questions below.

1. Name the type of weather system shown on charts (a) and (b).
2. Explain what the line with the half circles along it is meant to represent (Fig. 5.24a).
3. What is the difference between this line and the line with triangles?
4. Write down the wind direction you would expect at stations X and Y and state whether you would expect light or strong winds in each case.
5. Explain why the symbol for rain is shown at the station D.
6. Give reasons to explain the weather reported at station G.
7. Explain why there is a difference in temperature between stations D, E, and F.
8. Describe the weather you would expect to experience at station X.

SUMMARY: TYPES OF WEATHER IN MID-LATITUDES

Depressions (low-pressure regions) have upward-moving air which spirals anticlockwise and inwards in the Northern Hemisphere. They drag air in from all directions, but the different types of air take a long time to mix. Because air from the Equator is warmer and less dense, it rises over cold air from the pole along inclined planes called **fronts**. Warm air at fronts rises and cools, producing great sheets of stratus or nimbostratus cloud and precipitation.

The leading edge of the warm air wedge is called a **warm front**, the trailing edge is called a **cold front**. Between them lies the **warm sector**. When the warm sector rises completely off the ground an **occluded front** forms, accompanied by a wide band of cloud.

The sequence of weather associated with a depression moving past an observer on the ground (Fig. 5.16) is:

(1) Cold air with thin cirrus cloud. Ring around the moon.
(2) Cloud becomes lower and thicker and is accompanied by a belt of precipitation.
(3) The warm front passes, wind changes direction and comes from a lower latitude direction, and the temperature rises. Cloud becomes more broken and sunny periods occur.
(4) The cold front arrives abruptly, the air turns colder, and the wind blows from a higher latitude direction once more. This is accompanied by a further belt of precipitation.
(5) Rain ceases and cloud becomes higher and thinner with only scattered showers.

Anticyclones are quiet interludes between the hectic and often stormy events of depressions. They have downward-moving air which, in the Northern Hemisphere, spirals clockwise and outwards from the centre. As the air moves downwards it warms, and all its moisture can be held as vapour. There is no chance of precipitation. Skies are usually clear, and winds are light because the subsiding air spreads out slowly. With clear skies summer anticyclonic weather is hot and sunny, but in winter, when the angle of the sun is low and daylight much shorter, less heat reaches the ground than is radiated to space, so anticyclones bring cold, frosty weather. Indeed the lowest air might cool so much over night that **radiation fog** forms.

Topic Anatomy of a cyclone

Tropical storms — hurricanes, typhoons, and cyclones — are features of the climates of the eastern shores of the Americas, Asia, and Australia. They form as small low pressure "cells" that bud off the equatorial low pressure belt.

To make a cyclone, mix one large ocean having a surface temperature above 27°C with one fairly small tropical low pressure system. Allow to simmer gently over the ocean for about four days gathering energy and moisture and sending towering cumulonimbus clouds into the sky. Stir the storms vigorously, using the rotation of the earth (Coriolis Effect).

Cyclone Tracy was one product of this recipe. At 9 am on 22 December 1974 it was identified as a cyclone (wind speed in excess of 120 kph) in the Arafura Sea about 250 km northeast of Darwin, Australia. The storm intensified as it quickly followed the usual westerly then poleward route. As Tracy turned to the southeast on Christmas Eve, it became clear that Darwin would experience the full force of its violence. Warnings were issued and people took what shelter they could. But by 6 am on Christmas morning over 50 people had been killed and more than 90% of buildings destroyed.

Figure 5.25

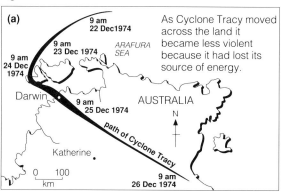

(a) As Cyclone Tracy moved across the land it became less violent because it had lost its source of energy.

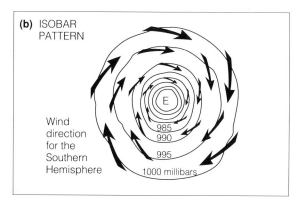

(b) ISOBAR PATTERN — Wind direction for the Southern Hemisphere

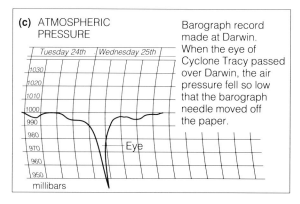

(c) ATMOSPHERIC PRESSURE — Barograph record made at Darwin. When the eye of Cyclone Tracy passed over Darwin, the air pressure fell so low that the barograph needle moved off the paper.

(d) WIND VELOCITY — Just before the eye of Cyclone Tracy arrived at Darwin Airport the anemometer was blown away. Estimated wind speeds.

Figure 5.26

Before Cyclone Tracy

After Cyclone Tracy

Exercise 5.6 Routes over the Atlantic

Figure 5.27 Plane routes over the Atlantic

One of today's major forms of travel is by plane. Millions of people fly from country to country for trade, a holiday, or to visit friends. One of the busiest routes is over the Atlantic Ocean between Europe and North America. With 10 000 L of fuel used per hour by the larger planes the most economical flight path must be chosen. However, this may not be the shortest distance because of a variety of weather conditions, as we shall see. For example it is important to fly through "thin" air where frictional resistance to movement is less, but at the same time it is vital to find a route either in the same direction as the global winds or at least a route with low head winds.

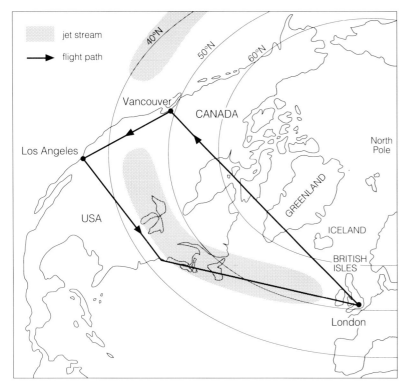

1. To obtain "thin" air would you fly at a high or low altitude (p. 57)?
2. In which direction do the main upper air winds blow over the North Atlantic (p. 56)?
3. Are there any special regions of high winds which could be a help (p. 56)?
4. Figure 5.27 shows the main type of air routes followed over the Atlantic to Los Angeles. Explain why different paths are chosen for the east and west directions.
5. Why are these routes changed daily after the air traffic controllers have received the daily weather forecasts?
6. In what way might these differences in wind be responsible for a flight from Los Angeles on the west coast of the United States to London taking 10 hours, but a flight in the opposite direction taking 11 hours?

Exercise 5.7 Geography behind the news

We can learn a lot about a region's geography by reading news stories. This news item is from *The Victoria Times-Colonist*, Sunday, 14 October 1984.

1. What kind of weather system must have developed over the northeast Pacific Ocean — a depression with a low pressure or an anticyclone with high pressure (p. 63)?
2. With winds gusting to 100 km/h, would the isobars have been close together or far apart?
3. The waves were remarkably high given the suddenness of the storm at Brooks Bay. What factors contribute to wave height?
4. Why do some fishermen and local residents want weather ships to be used again?
5. Why was this unexpected October storm particularly dangerous?

STORMY SEA YIELDS TWO BODIES

RESCUERS on Saturday found the bodies of two of four people thrown into the waters off the northwestern coast of Vancouver Island when their fishboats were swamped in a fierce storm Friday....

Search vessels ... found the two bodies, but no sign of the other two fishermen. Searchers hoped the pair had made it to land where they would have a better chance of survival.

Weather conditions off Brooks Bay Saturday were slightly better than on Friday ... but high seas and winds were still hampering the search.... A ground crew was also searching the coastline in the hope the fishermen were washed ashore.

The four had been in the water since early Friday morning when a storm with waves 10 m high and winds gusting to 100 knots savaged the coast. Seven vessels were swamped Friday and the Pacific salmon fleet scattered.

In Winter Harbour, the tiny fishing community closest to Brooks Bay and about 400 km northwest of Victoria, fishermen who had managed to make it to port huddled by themselves on their boats and blamed satellite weather forecasting for being caught unprepared by the storm....

Environment Canada scrapped its weather ships more than a year ago in favour of relying on satellite observation. Fishermen warned at the time that the change would cause problems as satellites can't tell wind speed.

6 The world's weather

Variations in weather

A global view

We have looked at key parts of the global weather machine and how it produces particularly complex weather patterns at latitudes similar to those of Canada. Now we shall consider how the global pattern affects other latitudes (Fig. 6.1).

In Chapter 5 we thought in terms of a simple pattern of air rising over the Equator. However, heating is not always greatest at the Equator: the zone of maximum heating moves as the overhead noon sun changes in latitude during the year. So the basic global circulation must include seasonal variations.

Seasonal changes in mid-latitude countries, such as Canada, involve slow temperature changes between summer and winter. There are no significant wet or dry seasons because depressions track across the area throughout the year. This type of "subtle" change is quite different from areas nearer the Equator. For example, the reason people vacation in Arizona, Florida, Mexico, or Hawaii is because they know that their winters are invariably warm, sunny, and dry. In these regions we find a fundamental difference between summer and winter, not only in temperature but also in precipitation.

However, it is one thing to have a dry season that brings in tourists and quite another to have to cope with a dry season if you rely on agriculture for survival.

Figure 6.1 Major climatic regions of the world

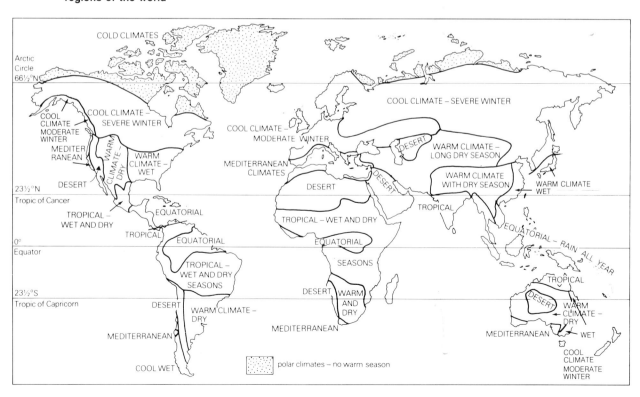

Topic The seasons

The earth spins on an axis which is tilted at 23½° to the plane of its orbit around the sun. The sun's radiation reaches the earth in parallel rays, making the lighted half of the earth *day* and the shaded half *night*. Because of the tilt, different places have different lengths of day and night as the earth completes an orbit around the sun each year. At the polar regions (beyond 66½°N or S), they receive 24 hours of daylight in summer and 24 hours of darkness in winter (Fig. 6.2). The time during each 24 hour period in daylight and the angle of the sun's rays to the surface determine the amount of solar radiation received to heat up the surface at any place.

Seasons are caused by the differences in day length and the angle to the overhead sun, and not by the different distances between earth and sun throughout the year. Notice that the tilt of the earth causes the Northern Hemisphere to receive more overhead sun in June (summer) while the reverse is true for December (winter).

We call the longest and shortest days of the year the **solstices**; they occur on about 22 June and 22 December. Days of equal periods of light and darkness are the **equinoxes**; they occur on 22 March and 22 September. All areas polewards of latitude 66½° receive 24 hours daylight or 24 hours darkness during the solstice; everywhere on the earth receives 12 hours day and 12 hours night during the equinoxes.

Figure 6.2 The seasons

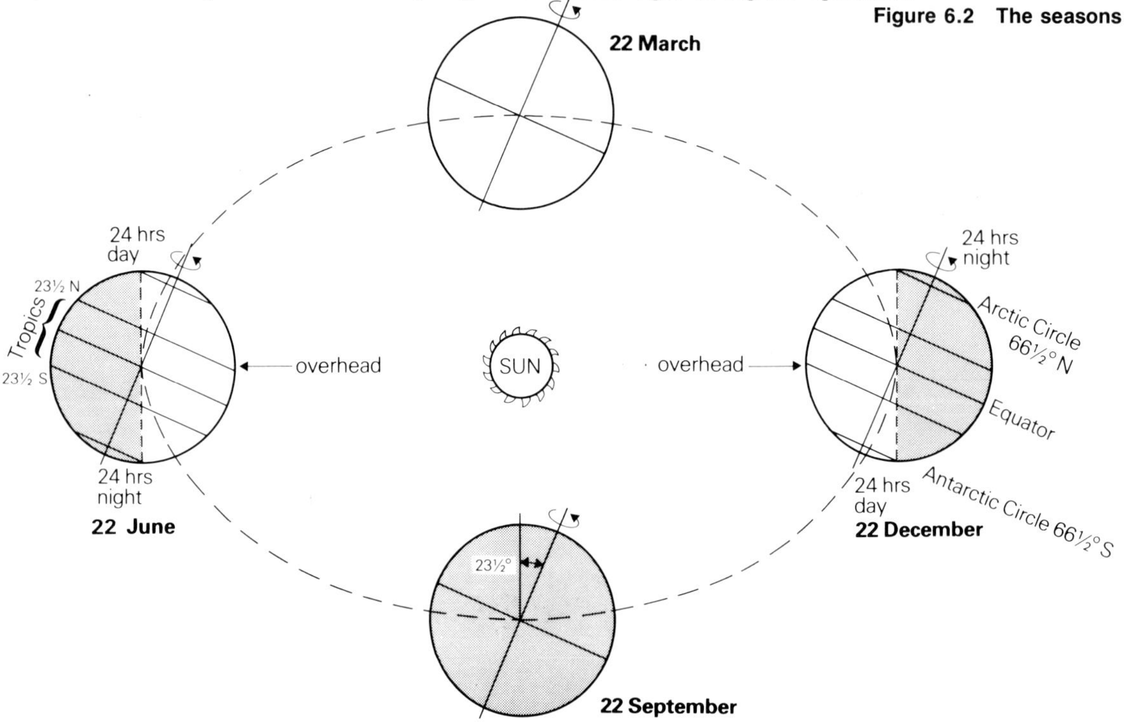

Exercise 6.1 British Columbia's climate

Although the mid-latitude depressions track a few degrees north and south during the year, western Canada remains under their domination. As a result, most of the winds are westerlies. By the time they reach BC these winds have been over many hundreds of kilometres of oceans, providing time for air temperatures to be modified. In Figure 6.3 a and b, air and sea temperatures are shown for winter and summer conditions.

1. Study the pattern of isotherms (lines of equal temperature). In winter is the sea warmer or colder than the land? Is this still true in summer?
2. In winter, what is the relationship between sea temperature, prevailing wind direction, and land temperature?
3. The winter isotherms for North America are lined up largely east and west, approximating the parallels of latitude. But summer temperatures are a great deal more variable, especially in the western half of the continent. When is the moderating influence of the Pacific Ocean most apparent with regards to average temperatures?
4. Why does the summer 20°C isotherm extend so far south in central and western USA?
5. What can you conclude from the pattern of precipitation along the west coast from California to Alaska?
6. What term would you use to explain the low precipitation in Alberta and the American Great Plains?

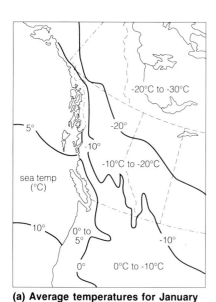

(a) Average temperatures for January

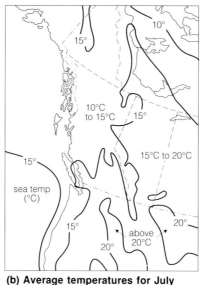

(b) Average temperatures for July

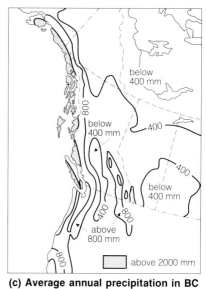

(c) Average annual precipitation in BC and Alberta

Figure 6.3 The climate of BC and Alberta

Exercise 6.2 Temperate continental and maritime climates

We normally describe regions often influenced by onshore ocean winds as **maritime.** This is because the ocean has moderated the air temperature, making it either slightly cooler in summer or warmer in winter. As a result, **maritime climates** do not experience as large a range of annual temperature as do continental interiors, which are far from oceanic influences. Similarly, precipitation is usually plentiful and evenly distributed throughout the year. **Continental climates** occur when the moderating oceanic effects are lost.

1. Study the climatic data for Prince Rupert, Prince George (British Columbia), and Edmonton (Alberta) (Fig. 6.4).
2. Use your atlas to make a rough sketch map of BC and Alberta. Mark the 3 places on your map.
3. Now write the total precipitation value beside each station. Draw arrows between the stations from the largest to the smallest. Can you suggest an explanation for the observed variation?
4. Measure the annual range of temperature (maximum minus minimum) for each station. Record these values on the map and draw arrows as before. Why does the range increase towards the interior of the continent?
5. State the main differences in precipitation distribution between the three stations.
6. Briefly summarize what you think are the main characteristics of (a) a maritime climate and (b) a continental climate.

Figure 6.4 Climatic data Prince Rupert, Prince George, and Edmonton

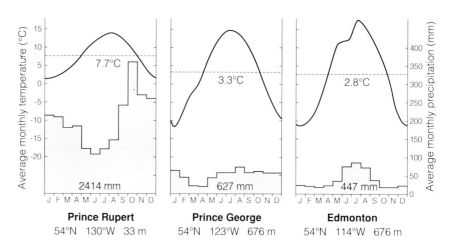

Prince Rupert
54°N   130°W   33 m

Prince George
54°N   123°W   676 m

Edmonton
54°N   114°W   676 m

In some tropical countries (those within about 20° of the Equator) a dry season can be entirely rainless with temperatures continuously above 30°C. This has a very significant effect on vegetation and the growth rate of plants, and on the numbers of animals that can be kept or the types of crops that can be grown.

Some tropical areas have a gradual transition from dry season to wet season; others suddenly experience a change overnight. These areas have a **monsoon** climate. At the end of the dry season the ground is parched and dry, with soils cracked open, and grass brown and dormant. Then one day rain falls so fast it cannot be absorbed by the soil and it runs in brown streams across fields and roads. This is the start of the wet monsoon season, when rain falls virtually every day.

January
(Northern Winter, Southern Summer)

Seasonal changes in climate

Follow the sun

Look at Figure 6.5 which shows two satellite photographs of the global circulation. Where air rises, cloud and rain are found; where air sinks, clear skies and drought prevail.

For a place X at about 10°N the change is dramatic throughout the year. In January X is under a sinking air flow and drought prevails, but in July

Figure 6.5 The apparent movement of the overhead sun

July
(Northern Summer, Southern Winter)

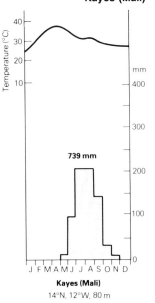

Figure 6.6 Climatic data for Kayes (Mali)

Kayes (Mali)
14°N, 12°W, 80 m
739 mm

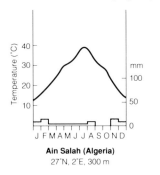

Figure 6.7 Climatic data for Ain Salah (Algeria)

Ain Salah (Algeria)
27°N, 2°E, 300 m

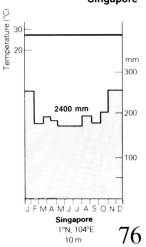

Figure 6.8 Climatic data for Singapore

Singapore
1°N, 104°E
10 m

2400 mm

it is under a rising air flow and it receives rain. As a result station X has a wet season and a dry season. It is typical of much of the tropical world (Fig. 6.6).

Station Z, at about 25°N, never emerges from within a region of sinking air. This gives it a **desert climate** (Fig. 6.7). Station Y, at the Equator, is under rising air throughout the year, so rain falls in every month. This is characteristic of an **equatorial climate** (Fig. 6.8) and it provides the moisture for the rain forests of Southeast Asia, South America, and Central Africa.

The variation in pressure systems from January to July is clearly connected with the movement of the overhead noon sun. On 22 June (summer solstice) the sun is overhead at midday at 23½°N of the Equator, bringing the zone of greatest heating with it. The coldest northern regions are polewards of about 70°N. As a result, all the patterns of air flow in the Northern Hemisphere are shifted north and crushed up between 20°N and 70°N. It is this movement that provides the rainy season to the Northern Hemisphere tropics.

The polar high breaks down for a brief period during June, July, and August, and the Arctic region experiences its fleeting summer. Meanwhile, the subtropical high which helps to create the deserts of Mexico and the southwestern USA expands northwards to encompass most of southern BC, Alberta, and Saskatchewan. Offshore winds prevail.

On 22 December (winter solstice) the sun is overhead at midday 23½°S of the Equator, taking the zone of greatest heating with it. As a result the pressure belts in the Northern Hemisphere expand southwards, with the systems near the overhead sun moving the greatest amount. Thus, in the tropics, what were rainy areas in June come under the influence of sinking air, and the dry season spreads to Mexico and the northern tropics. As a result the wet season begins in the southern tropics.

Of course, it is not surprising that all of the world's climates are linked together in this way. Basically climates exist in latitudinal belts or *zones* around the earth. This is clearer in the Southern Hemisphere (which is mostly water) than in the Northern Hemisphere where large land masses make the pattern more confused.

Hot stuff! *The influence of land masses*

The main difference between oceanic and continental heating has already been discussed. Water absorbs and stores heat in its surface layers. As a result, water temperatures do not vary much all year in the oceans. By contrast, the land conducts heat very poorly; its surface heats up considerably in summer and cools down dramatically in winter. Heated surfaces well away from the sea cause air in contact with them to warm and rise, thus creating low pressure regions; the reverse is true for cooled surfaces, which create high pressure regions. Indeed the major land masses are so big that the effects of their heating and cooling are very influential and often outweigh the effects of the main global circulation patterns. Only the subtropical high pressure belt is strong enough to overcome the effects of continental heating. As a result, deserts are found even though surface temperatures are very high.

Asia, the largest land mass, is the most important land area distorting the global picture in this way, producing an intense low pressure system in summer and a high pressure system in winter. The effect of the summer low is to draw air in over the continent from nearby oceans. Some of this air has enough moisture to bring considerable quantities of precipitation with it, often in the form of heavy and prolonged downpours. By contrast, in winter pressure becomes high with subsiding and outflowing air as the continent cools down quickly. As a result, most of the moisture-laden air is blocked out, and a dry season prevails.

Upper-air winds

It's an ill wind...

It is not possible to separate the patterns of weather on the ground from the movements of air high in the sky. Nowhere is this more true than in monsoon lands.

Monsoons are characteristic of some areas with distinct wet and dry seasons (Fig. 6.9). In such areas the wet season arrives very suddenly. For example, the leading edge of the summer monsoon in India usually reaches the southern tip of the country on about 15 May, after a couple of months without any precipitation at all. It is so regular and it starts so suddenly that people refer to the monsoon as being early or late if it is only a few days different from normal. This suddenness is surely not explained just by gradual seasonal shifts in the global circulations; some other force must be involved.

In India, for example, the effect of a jet stream on the monsoon is very important. In December the Northern Hemisphere monsoon jet stream flows at its nearest to the Equator (Fig. 6.10). This brings it over northern India, south of the Himalayas. At this time of year a high pressure develops and the dry season prevails. In the spring the air patterns over the earth all start to move northwards as described earlier. But over India this movement is hampered by the Himalayas and the jet stream appears to get stuck to the south of the mountains. As the summer approaches, this jet stream is farther and farther out of place until it finally "flips" and suddenly moves to the north of the mountain range. Although we are not sure precisely how this happens, the result is certainly dramatic. One day there is a jet stream containing lots of air; the next day it is gone and there is a great lack of air over northern India.

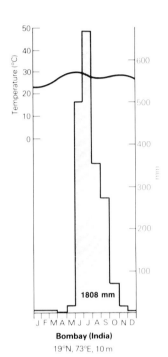

Figure 6.9 Climatic data for Bombay (India)

Figure 6.10 The causes of the monsoon
The rainy season is due to summer heating over India causing a low pressure region and pulling moist air in from the ocean. The abrupt start to the monsoon is caused by an upper air movement of a jet stream.

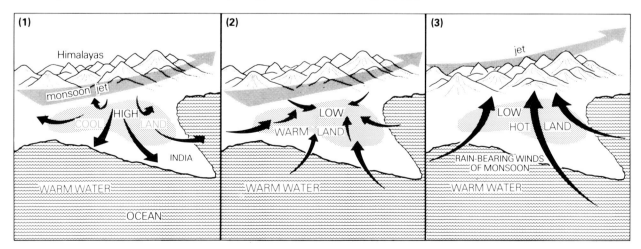

Moist (rain-producing) air now rushes in from the Indian Ocean to fill up the space left by the jet and in doing so starts the monsoon. Once started the normal summer heat keeps air rising over the land and drags more air and precipitation in to prolong the rainy season.

Polar climates

Real cool!

The complex tropical and subtropical lands lie to one side of the mid-latitude "gears" and occupy more than half the earth's surface. The cold lands lie to the other side, but their extent is very small, being less than 10% of the earth's surface. Here there are fewer than four months with temperatures above 10°C

Exercise 6.3 Tropical climates

Figures 6.6, 6.8, and 6.9 show climatic information for three places in the tropics.

1. Describe the pattern of temperature at the three stations.
2. Explain why Bombay and Kayes (Mali) have a much larger range of temperature than Singapore (look carefully at the latitude of each place).
3. What is the name given to the abrupt change from dry season to wet season shown in Figure 6.9?
4. Explain why many tropical places have a wet and dry season.

Exercise 6.4 Desert climates

1. Study Figure 6.7. Hot deserts are often defined as having less than 250 mm of precipitation a year. Add up the precipitation bars. Does Ain Salah come within this classification?
2. With a very low precipitation how much cloud might this area experience?
3. What pressure belt influences this area and keeps precipitation low?
4. Temperatures vary widely between summer and winter. Why should this be so?
5. Deserts can be sustained by high pressure zones, rainshadows, or simply distance from the sea. Make a rough copy of Figure 6.11 and indicate the probable cause of each of the deserts marked.

Figure 6.11 The world's deserts and areas liable to desertification

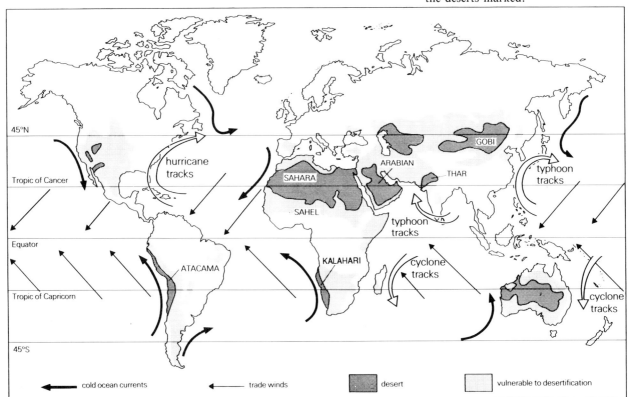

and the weather is dominated by cold winds from the poles. Nevertheless in these regions, where long periods of clear skies prevail because of the domination of sinking polar air, there is a great range of temperature. In winter, −30°C would not be an unusual temperature when darkness occupies much of the 24-hour period and the sun only nudges above the horizon. In contrast, temperatures on some summer days can rise to 20°C because of the long hours of sunshine. Tomsk (Fig. 6.12) lies at the southern limit of this area, to the north conditions become even more severe. Real polar areas have no month with an average temperature above 10°C, so their temperatures range from this value downwards. The lowest temperature recorded at the South Pole is −150°C!

Exercise 6.5 Cold climates

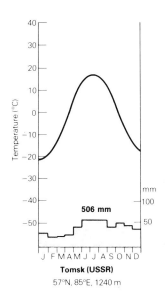

Figure 6.12 Climatic data for Tomsk (USSR)

Cold climates have very few months above freezing and therefore a short growing season. Tomsk, in Siberia, (Fig. 6.12) is in the coniferous forest belt of Asia, but only a little way north conditions become too severe for trees to grow, and tundra conditions begin (see p. 132).

1. If plants require a sustained temperature of over 6°C for growth, how long is the growing season in Tomsk?
2. For how many months is it likely that precipitation will fall as snow?
3. For how many months is the temperature above freezing?
4. Find Tomsk in an atlas. From its position in the Asian continent can you suggest why it has such a low total precipitation?
5. Climatic conditions such as these must play a large part in determining the type of houses and the lifestyle of the people. Suggest what some of the main problems must be and the way they might be overcome.

SUMMARY: THE WORLD'S CLIMATES

Hot climates. Hot climates have a coldest month above 18°C and they occupy a broad belt well beyond the tropics, thereby including over half the earth's surface. Seasonal and daily changes in temperature are controlled by the movement of the overhead sun. This gives two maxima of temperature (spring/autumn) and precipitation in equatorial areas and single summer temperature/precipitation maxima in other tropical areas.

(a) *Equatorial:* With the sun overhead in spring and autumn there is no summer maximum and fairly uniform temperatures prevail all the year. Precipitation occurs in all months but is heaviest in spring and autumn just after the sun has passed overhead.

(b) *Tropical:* These regions lie at the edges of the equatorial rain belt. They experience dry trade winds for most of the year, with rain bearing winds at other times. Many eastern coastal margins experience **hurricanes** and **monsoon** precipitation. There is considerable temperature range through the year, with 27°C being a common maximum for the hottest month.

Warm temperate climates. Warm temperate climates lie between the trade winds and the westerlies and they are transition climates, having a summer "tropical" season and winter similar to temperate areas.

(a) *Mediterranean* (western margin): These are coastal climates, with hot dry summers (over 20°C in warmest month) and wet mild winters (above 6°C in coolest month).

(b) *Eastern margin:* These areas have mild winters and hot summers but the precipitation patterns are reversed. Low winter rainfall and plentiful summer rains are the rule, but there is no long dry season as with Mediterranean-type areas.

Cool temperate climates. Cool temperate climates have a cold season when plants do not grow. Seasonal changes in temperature are more important than changes in precipitation. These regions are dominated by weather from depressions which track along in the belt of westerly winds. Continental areas have a large seasonal temperature range. Precipitation occurs in all seasons but decreases away from the coasts; interior regions have a summer maximum because convection precipitation is most important.

Cold climates. Cold climates are dominated by depressions and anticyclones in the same way as cool temperate regions, but as winters at higher latitudes are longer and more severe, most outdoor activity is restricted.

Desert climates. These climates occur in diverse regions where temperature and precipitation vary widely. Often annual precipitation below 250 mm coincides with these areas, but a scant vegetation is the best indicator of aridity.

(a) *Hot deserts* do not have cold seasons below 6°C. They occur in subtropical high-pressure regions and are reinforced at coasts by cold, upwelling ocean currents.

(b) *Cold deserts* have at least one month below 6°C and are mostly in rainshadow areas or at the hearts of continents.

Climatic change **...and what is normal for this time of year?"**

Farming has always been a hazardous business in the southern Canadian Prairies. In many areas, precipitation has been low and unreliable for most of the past 150 years, and farmers have relied on the occasional good season to carry them through the dry times. However, during the 1980s the precipitation has been even more meagre, and in 1987-1988 the Prairies had its fourth drought in a decade. Since 1981, precipitation has declined to about 50% of long-term averages. Not surprisingly, in 4 of the past 10 years, high temperature records have been set. At Lethbridge, Alberta, April 1988 became the driest month on record as no measurable precipitation was received. Lethbridge is in the heart of the Paliser Triangle, the hardest hit region on the Prairies. Lack of winter snow and, therefore, moisture has meant seeds have not germinated. With little vegetation to bind the topsoil, hot, dry winds have carried the topsoil away and evaporated water supplies. Many towns and cities, faced with acute water shortages, enforced stringent water-rationing. At Assiniboia, a small town in southern Saskatchewan, people who watered their lawns lost their water supply for 48 hours and were fined $500! But nature seemed to be at its worst when the choking, blinding dust storms occurred. Sometimes they were so severe drivers had to use their headlights at midday.

Cattle ranchers were the first to feel the adverse effects of the drought. Severe damage to Prairie grasses and soil might mean that ranching will be the last agricultural activity to recover from the drought.

These climatic changes might be part of normal, continuing trends — a first, obvious sign of the greenhouse effect. Within 50 years, the growing season of the Prairies might increase by 40% to 140 days. But droughts will probably be more frequent and more severe. Farming and ranching activities might, therefore, move or change to more closely match weather and climate.

Exercise 6.6 Mediterranean climates

In the table below, the temperature and precipitation data are given for Marseilles in southern France (43°N, 5°E).

1. Use the figures below to draw a bar graph similar to all the others in this chapter.
2. A Mediterranean climate has (a) most of its precipitation in the winter with very little in summer; (b) hot summers with the warmest month above 20°C and mild winters with the coldest month above 6°C; and (c) much sunshine, especially in summer. Show how Marseilles fits into a Mediterranean type of climate.
3. State how you can tell it is sunny from the information in the table.
4. The minimum temperature for plant growth is 6°C. The Mediterranean coast of France, especially the area to the west of the River Rhône, is well known for its vegetables. At what season of the year do you think growers try to sell their crops in order to get maximum profits from the markets of northern Europe?

	J	F	M	A	M	J	J	A	S	O	N	D	
Temperature (°C)	7	8	10	13	16	20	22	21	19	15	11	8	
Rainfall (mm)	43	36	48	56	43	28	18	20	61	96	71	53	Total 573

7 The waters of the oceans

Oceanic circulation

Swirling waters

When large storms develop in the open ocean they can hurl water about in tumultuous waves up to 20 m and 30 m high; they can even break the back of a supertanker. But observed on a global scale, even these giant waves are nothing more than surface ripples driven by the friction of winds on the ocean surface. The bulk of the ocean water remains completely unaffected by the surface waves — it is involved not in these local disturbances but in a world-wide transfer of energy. Only in shallow seas does the friction of wind on the water reach down to the sea bed.

The movement of large volumes of water is just as important to the world's heat balance as the movements that take place in the atmosphere. Two thirds of the earth's surface is water and thus receives two thirds of the sun's energy reaching our planet. However, water absorbs this energy quite differently from land. Some times, such as when the sun's rays strike the water surface obliquely, virtually all the energy is reflected back to space (Fig. 7.1). However, when the sun's rays come from overhead, 90% of the energy is absorbed by the surface layers of water, providing light and warmth for millions of sea creatures and also a vast store of heat for the earth.

Notice that it is the upper layers of the ocean that absorb the sun's radiation and warm up. Warm water is less dense than cold, so the heated surface waters stay on top of the oceans. This is very different from atmospheric heating, in which air is warmed by contact with land or sea and so is heated from below. With air warmer near the ground and colder at height, the warm air rises and produces all sorts of turbulent effects in the atmosphere. With warm water on top, oceans are rarely affected by this sort of turbulence.

Although oceans are heated from above and the atmosphere from below, both receive much more heat near the Equator than near the poles. As a result oceans are warmest in equatorial regions and coldest near the poles. As with the air, these temperature contrasts cause worldwide convection currents to

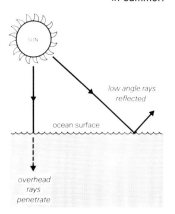

Figure 7.1 Heating the ocean
The ocean absorbs heat when the sun is overhead, but very little when it strikes at a shallow angle. Arctic oceans receive very little heating even though the sun shines 24 hours a day in summer.

Topic Why the sea is salty

When a volcano erupts it throws out great volumes of steaming water along with gases, ashes, and lava. Water in the oceans, on land, and in the atmosphere may have originally come from volcanoes in this way. New water is still being added to the earth's surface by the same means.

Every new addition of water contains a great variety of minerals dissolved in it, but somehow the oceans are dominated by just common salt (sodium chloride, NaCl). Similarly rivers contain a great variety of minerals in solution eroded from the surface of rocks. Yet the flow of river water in the oceans does not upset the salt concentration at all.

Most of the minerals carried into the oceans, from either volcanoes or rivers, are used to make the shells of sea creatures or else are absorbed by freshly settling sediments on the ocean floor. Some minerals such as manganese even manage to concentrate themselves into nodules which are left to litter the sea bed. But sodium and chlorine are not completely used up in this way and so are left as the main mineral components of sea water. When water evaporates to form moisture in the air, the salt is mostly left behind. Any salt that gets into the air as spray from breaking waves is washed out by rain and returned to the sea by rivers. It is this long-term balance that keeps the sea salty.

form in an attempt to spread out the heat energy more evenly (Fig. 7.2).

We see the ocean convection as huge volumes of water constantly on the move. These are called **ocean currents** and many of the largest have been given names by the sailors who make constant use of them. Names such as the **Gulf Stream** are familiar to most people. This is a large volume of warm water that travels from the Caribbean Sea along the east coast of the United States and finally crosses the Atlantic to provide warmth to the shores of Europe.

However, warm surface waters such as the Gulf Stream are only part of the worldwide circulation. They must be balanced by returning currents of cold water. Warm currents are usually fast-moving, relatively narrow bands of water; cold currents move much more slowly, but spread out more.

Figure 7.2 The main ocean currents of the world

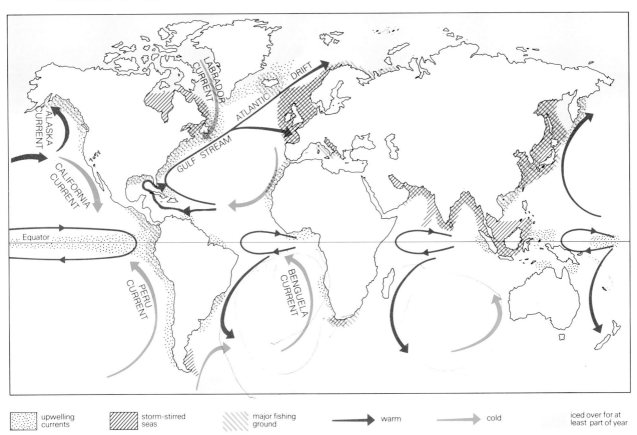

Exercise 7.1 Ocean currents

1. Look at Figure 7.2 and concentrate on the Southern Hemisphere first. Describe any current pattern you see in the South Atlantic Ocean. Can you find the same type of pattern in the other southern oceans?
2. Now compare the northern and southern Atlantic ocean currents. What is the basic difference?
3. What happens to the ocean flow near the Equator?
4. Copy and complete Figure 7.3, which is a simple drawing of the ocean currents. Notice each ocean has a *figure of eight* pattern of currents. This may help you remember the ocean currents more easily.
5. Thor Heyerdahl, an adventurer and explorer, developed theories about population migrations and ocean currents. Read about his discoveries and write a brief report.

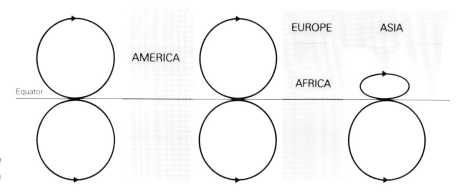

Figure 7.3 A simple model of ocean circulation

Wind and convection

Current affairs

Because the oceans are warmed by the sun's energy more near the Equator than near the poles, large-scale convection currents develop. The currents that even out the oceans' heat are often well defined "threads" of water. Those bringing warm water to the polar regions are called **warm ocean currents,** while those bringing cold water to the Equator are **cold ocean currents.** Convection currents cause the main north-south flows, but many of the east-west flows are partly dragged along by winds such as the trades and westerlies.

Convection and wind drag effects are clearly seen, for example, in the North Atlantic Ocean. Here, off the coast of Venezuela, the trade winds blow so constantly from northeast to southwest that they drag the surface water from Africa westwards until it piles up in the Gulf of Mexico (Fig. 7.2). All the while the water is heated by the sun and it gradually becomes warmer and warmer. But water cannot collect in the Gulf of Mexico indefinitely, so some of it pours continuously out to the north as the trade winds drag more in from the east. You can form some idea of the enormous volume of water involved when you know that the Gulf water is 20 cm higher than the Atlantic. The sea level actually slopes down to the northeast allowing gravity to help water to pour out. About 26 million m^3 of sea water flow from the Gulf to the Atlantic every second. This is the water that starts the famous Gulf Stream.

Convection currents help transfer water towards the poles and create a return current of cold water southwards along the coast of Africa. As the warm water moves north it helps warm up the air in contact with it. This warmed air then flows over northwest Europe to give an exceptionally mild climate for those latitudes.

Fronts in the ocean

Icy fingers

Most of the ocean currents are quite uniform over large distances, but problems come when water that is cold at the surface meets warm surface water. The Arctic Ocean is always cold, from the surface down. Indeed, because water is most dense at 4°C, the deepest part of the Arctic is actually warmer than the surface. Because it is so cold, the surface of the Arctic Ocean is often a mass of pack ice and icebergs (Fig. 7.4).

The cold Arctic water is constantly drawn southwards as part of the worldwide convection in the oceans. With it comes a generous surface coating of broken **pack ice** together with **icebergs** that have broken away from the glaciers around the coast of Greenland. This cold water comes farther and farther south until, somewhere off the coast of Newfoundland (the Grand Banks), it meets the warm Gulf Stream. With cold and warm water both on the surface, mixing occurs in an irregular way like the interlocking fingers of two hands — one a set of icy fingers stretching out southwards, the other a set of warm fingers pushing their way northwards.

Eventually the warm Gulf Stream water ends up on top of the cold Arctic water, and the broken pack ice and the icebergs melt. It is this region of mixing that has been a constant terror to generations of North Atlantic seafarers (see also fog, p. 65). On Wednesday 10 April 1912 the newly built ocean liner *Titanic* sailed on its maiden voyage from Southampton, England to New York. At 60 000 t it was the largest ship then afloat, rising 25 m out of the sea and extending 300 m from bow to stern. At just after midnight on a calm, starlit night the ship's engines were switched off and the huge vessel came quietly to rest, not 48 hours steaming time from New York.

Full of curiosity, many of the 2208 passengers and crew came up on deck to find out why the giant ship had stopped. Most were unaware that, a quarter of an hour previously, the ship had struck the submerged part of an iceberg and tore a 90-m gash in its right side. There had been only a slight lurch and a faint scraping noise. A passenger later reported "Suddenly... a hissing roar... made us all turn (to see) a rocket leap upwards... and then an explosion... and a shower of stars sank slowly down and went out one by one. And with a gasping sigh one word escaped the lips of the crowd: 'Rockets'! Anybody knows what rockets at sea mean!"

With water pouring into 5 of the 16 "watertight" compartments, a full-scale evacuation took place and by 2.15 am the ship had almost sunk.

"And then, as we gazed awe-struck, she [*sic*] tilted slowly up, revolving apparently about a centre of gravity just astern of midships, until she attained a vertically upright position... then she slid slowly forward... and the sea closed over her."

Figure 7.4 Pack ice and icebergs
Left. Pack ice breaks up into fragments. *Right.* These icebergs, off the coast of Newfoundland, calved from glaciers in Greenland. As they drift southwards on the Labrador current, the icebergs present a danger to shipping and offshore rigs.

Topic The reasons for tides

Movements of ocean currents and atmosphere are mainly driven by energy from the sun, but tides are most strongly influenced by the gravitational pull of the moon.

Everything in the universe has a gravitational field and attracts and is attracted to everything else. The strength of the gravitational pull depends on the size of the body and its distance from those it is attracting. So the larger the body, the greater the pull — at least as far as our everyday experience is concerned. Because the gravitational pull by the moon is less than that of the earth, for example, astronauts have been able to make very high jumps on the moon. We, however, are used to everything being firmly anchored by the earth's gravitational pull. Indeed, if it weren't for gravity the spin of the earth would throw everything off into space. Yet although the main force on the earth is our own planet's gravity, the pull of the moon is still great enough to attract the ocean waters (Fig. 7.5).

When the moon is overhead the moon's gravitational pull drags water towards it. We call the piled-up water a **high tide**. As the moon goes round the earth so the high tide goes with it; and as the moon encircles the earth every 28 days but the earth spins on its axis every 24 hours, the earth and moon get out of step by 52 minutes a day. As a result, high tide is 52 minutes later each day (Fig. 7.6).

There are two high tides every lunar day. The other high tide occurs exactly opposite that caused by the moon's attraction, but is explained by the very peculiar way the earth and moon revolve together. It does not necessarily produce the same height of tide as that due to the moon's attraction.

The pull of the sun is much less than the moon because the sun is so much farther away. But when the sun and moon pull in the same direction a very high tide occurs — a **spring tide**; conversely, when the pull of the sun is at right angles to the pull of the moon, water is pulled from two directions at once and a low high-tide occurs — a **neap tide**. Neap and spring tides occur alternately about every 14 days.

If the moon drags a high tidal flow of water across the earth's surface and into a funnel-shaped piece of coast, the tides will become higher and higher towards the narrow end of the funnel. This is why the coast of the Bay of Fundy has such high tidal ranges. These high tides can push water back up river channels for many kilometres. One of the most famous examples is the **bore**, or tidal wave, which pushes up Chignecto Bay and the Petticodiac River in New Brunswick.

Tides can be an advantage or a disadvantage to people. A high tidal range is used to drive turbines and produce electricity in the Rance estuary in Brittany, France. Here the estuary has been dammed and the water flows in and out through turbines (Fig. 7.7).

However, the same high tidal range presents great difficulties for shipping. London, Rotterdam, and Le Havre are examples of ports which large ships can enter only at high tide. But Marseilles is a Mediterranean port where tides are very small and ships can enter at any time. This is a major economic advantage because ships waiting out at sea for a high tide are not earning money.

The growth of Vancouver has also been affected by tides because Vancouver Island causes ocean water to reach the Strait of Georgia by two different routes. Fishing boats, tugs with log booms or strings of barges, cruise ships, and cargo vessels all use the tidal currents to their advantage.

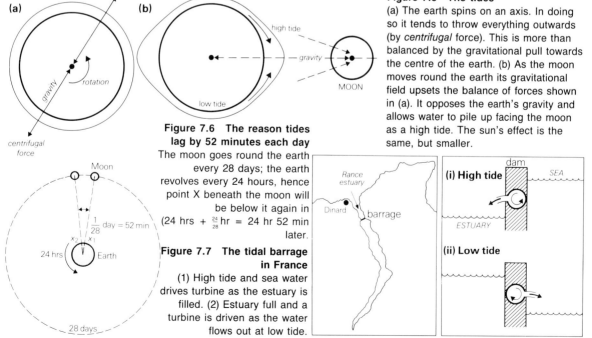

Figure 7.5 The tides
(a) The earth spins on an axis. In doing so it tends to throw everything outwards (by *centrifugal* force). This is more than balanced by the gravitational pull towards the centre of the earth. (b) As the moon moves round the earth its gravitational field upsets the balance of forces shown in (a). It opposes the earth's gravity and allows water to pile up facing the moon as a high tide. The sun's effect is the same, but smaller.

Figure 7.6 The reason tides lag by 52 minutes each day
The moon goes round the earth every 28 days; the earth revolves every 24 hours, hence point X beneath the moon will be below it again in (24 hrs + $\frac{24}{28}$ hr = 24 hr 52 min later.

Figure 7.7 The tidal barrage in France
(1) High tide and sea water drives turbine as the estuary is filled. (2) Estuary full and a turbine is driven as the water flows out at low tide.

At the time of the *Titanic* disaster (in which 1500 people drowned) the only way to detect an iceberg was to mount a lookout (with a telescope) at the bow of the ship. Radar was not invented until 30 years later. It was not even possible to detect the presence of icebergs accurately by observing sea temperatures. Because this is a region of mixing, temperatures can vary from 1°C to 14°C within a kilometre or two and icebergs can easily float south for hundreds of kilometres beyond the mixing zone before being melted away completely. The *Titanic* disaster happened at latitude 42°N, about the same latitude as the California-Oregon state border.

Global fishing grounds

Where shall we fish tomorrow?

Commercial fishing is the most important economic activity along our Pacific and Atlantic coasts. New and varied techniques are now used to harvest herring, salmon, tuna, bottom fish, crustaceans, and shellfish. Technological innovations such as radar, depth sounders, fish farms, and drift nets have reduced some of the hazards and uncertainties of fishing, but many new technologies wreak havoc by killing all species of sea life indiscriminately.

Before technological advances, people had to rely solely upon their knowledge of ocean movements and the biological life cycle to know where to fish. Ocean movements and the life cycle overlap as we can see by following the fate of a fish that dies. Burial at sea occurs in a relentless way as soon as the dead creature sinks to the seabed. Here minute particles of sediment are constantly settling out of the water and building up on the sea floor (Fig. 7.8). Eventually they become sedimentary rock. If the rate of sediment buildup is fast enough, the dead body will be encased in the sedimentary ooze and then be subject to a slow "embalming" process which will eventually turn it into a fossil. However, the seabed is usually a very active place, for it is the home of millions of scavenging creatures whose entire existence depends on finding food to eat. Much of this work is achieved by bacteria, which, as part of their "eating" process, release large quantities of nutrients into the sea water. But these are the very nutrients that are needed urgently back in the warmer waters near the surface to allow the growth of plankton. **Plankton** are microscopic sea creatures that float in the upper waters of the oceans. They are the main food of most fish. As a result, continued survival of plankton, fish, and scavengers depends on getting the nutrients from the seabed back to the surface. With warm water on the surface anyway, there is no convection mechanism to overturn the water, stir up the bottom, and bring the nutrients to the surface, so an external force is needed to stir up the water.

Figure 7.8 Storm turbulence Winter storm winds stir up shallow seas so that nutrients are brought off the seabed and returned to the surface.

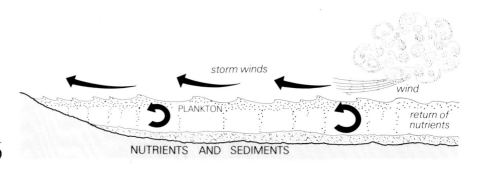

Fishing in shallow seas

When the north wind blows

With no internal mechanism, the stirring can occur only through the help of some external force. In fact two main forces operate: the wind and mixing ocean currents, the first only in shallow seas and the second in either seas or oceans. So the distribution of plankton concentrations, fish, and the world's fishing industry depend on finding places where the effects of winds can reach the sea floor or where ocean currents mix.

The Grand Banks provides an example in which strong winds reach to the seabed. The strong winds that blow over the Grand Banks in winter are part of the deep depressions that send whirlpools of air across the Maritimes. As they blow they transfer some of their energy to the ocean by friction, stirring it up and creating waves (Fig. 7.8). But stirring the surface of the shallow Grand Banks can produce a "stirring" effect right down through the water to the seabed. The winds and the cold Labrador ocean current mixing with the warm Gulf Stream produces turbulent mixing which drags the nutrients from the continental shelf to the warm surface waters (Fig. 7.10). In winter this gives the Grand Banks its greeny-brown and cloudy look, because sediment and nutrients are all stirred up together. But in the spring, when the winter gales die down and the sun warms the surface again, the nutrients are still there in suspension and can provide the food for plankton. As the plankton multiply (called "blooming"), their presence attracts the hungry shoals of fish. Of course some fish short-circuit this food chain and feed directly from the bottom, but they can do this only in shallow, nutrient-rich waters. Thus seabed turbulence produced by strong winds provides the basic cause of the Grand Banks' prolific fishing grounds, alas now suffering from overexploitation. Grand Banks winds blow from many different directions and stir up the seabed by force, but even a gentle wind can drag sea water along and stir up the bottom of a shallow sea if it blows constantly for long periods of the year. Thus the coastal regions of Arabia and the Bay of Bengal are also important places for fishing, even though they do not receive many storms. (Fig. 7.9).

Deep-sea fishing

Fruits of the sea

In just the same way that there are frontal zones which separate regions of warm air from regions of cold air, so there are frontal zones in oceans separating regions of warm water from regions of cold. These are places where two masses

Figure 7.9 Offshore winds When offshore winds blow constantly they push the warm surface water away from coasts. Cold, dense water from the depths rises to replace the warm water and in doing so drags up nutrients from the seabed. In this way the basis for plankton growth is provided.

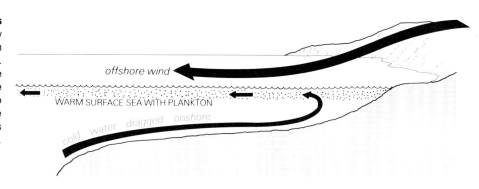

of water collide and struggle past each other, usually in an irregular way as we saw in the *Titanic* disaster. Where two strong currents meet there can even be regions of rough sea. Where two currents collide all sorts of floating particles are concentrated, including oil slicks. But, of special interest to fishing crews, they are also places where plankton become concentrated and which, therefore, attract fish. As a result, ocean fronts are some of the world's most productive fishing grounds and their study is of vital importance.

The fronts can occur in the middle of oceans, as Thor Heyerdahl discovered during his Kon-Tiki expedition across the Pacific in 1947. His observations, although indirect, show the effects well. When he had drifted more than 1000 km off the coast of South America he encountered a region of very rough water. This was the turbulent zone where the fronts met.

> The crest of the sea reached two metres above the level of the roof of the bamboo cabin, and if two vigorous seas rushed together, they rose still higher in combat and flung up a hissing watery tower which might burst down in unexpected directions. Next day we noticed more life than usual in the sea... during the afternoon we saw a big fish approaching the raft close to the surface... When we were having a rather wet and salty midday meal... a large turtle was lifted up by a hissing sea right in front of our noses... (then)... it was gone as suddenly as it had appeared (and)... we saw the gleaming whitish-green of dolphins' bellies tumbling about in the water below. The area was unusually rich in tiny fish an inch long which sailed along in big shoals... At 6.20 pm (Erik) reported our position as latitude 6°42′S and longitude 99°42′W...

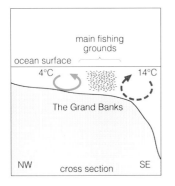

Figure 7.10 The Grand Banks fishing zone
This major fishing zone results from a meeting of cold and warm water. Inshore the sea-bed is also stirred up by storms.

Exercise 7.2 Fishing grounds

1. Copy the fishing grounds from Figure 7.2 onto an outline map.
2. Using Figure 5.10, indicate (on your outline map) the shallow sea areas which usually experience either westerly winds (blue) or trade winds (red). Check your atlas to confirm that seasonal wind patterns confirm your answer.
3. Why are some potential fishing grounds not exploited? (You will need to consider each location separately.)

8 The water cycle

Rainfall patterns

Feast and famine

Southern Manitoba is a remarkably flat area with rich alluvial soil. Its rivers form part of a great drainage system which extends east to Ontario, west to Alberta, and south to the Mississippi in South Dakota. This vast, gently sloping basin, only 255 m above sea level at Winnipeg, is drained by such major rivers as the Red, Assiniboine, and Saskatchewan. Although this area is occasionally flooded by springtime rain run-off and snow melt, it often experiences severe drought. Journals from early pioneers hint at the difficulties they faced because of too much or too little water.

26 October 1804 The canoes go no further up the river due to the shallow water this season. *(Journal of D.W. Harmon,* Fort Alexandria, Upper Assiniboine River)

1 June 1805 The river is very low as we have not had a drop of rain since last autumn. *(Journal of D.W. Harmon)*

13 August 1806 The summer's extraordinary rain, having overflowed the low country has caused the buffalo to resort to the highland southward. Famine general amongst the Indians.

4 May 1826 The water overflowed the banks of the river and spread so fast that almost before the people were aware of the danger it had reached their dwellings. *(Henry Journal)*

3 May 1852 Many settlers deserting houses due to rising water *(Notes of the Red River Flood* by Bishop Anderson)

12 February 1868 The crop of 1868 was a complete failure — even seed having to be imported (due to drought and grasshoppers). *(The Nor-Wester)*

Meteorological records for Winnipeg have been maintained since 1872, and have indicated variable precipitation and moisture conditions. Floods and droughts have repeatedly challenged people's resources. However, floods have caused the most obvious damage. Here are records of the five greatest floods of the Red River at Winnipeg:

DATE OF MAXIMUM DISCHARGE	ESTIMATED MAXIMUM DISCHARGE IN CUBIC METRES PER SECOND (m^3 ps)	PROBABLE RETURN PERIOD (YEARS)
21 May 1826	6 360 m^3 ps	667
21 May 1852	4 660 m^3 ps	147
8 May 1861	4 530 m^3 ps	45
10 May 1879	2 995 m^3 ps	26
19 May 1950	2 927 m^3 ps	23

The 1950 floodwaters covered about 1280 km^2, and the river spread to a width of about 24 km at Winnipeg. Flood damage, in 1989 values, was approximately $500 million. After the 1950 disaster a Royal Commission was formed to study the costs and benefits of flood control. The Commission's recommendations resulted in five major developments.

1 *Red River Floodway* A 46 km floodway channel bypasses Winnipeg. When the Red River flow exceeds 885 m^3 ps, the floodway handles the excess water, up to 1700 m^3 ps.

2 *Shellmouth Reservoir* A 21 m high and 1260 m long dam at the juncture of the Assiniboine and Shell rivers, has created a 480 million m³ reservoir which extends for 56 km. There is 280 million m³ of storage for floodwaters.
3 *Portage diversion* A diversion channel from Portage la Prairie to Lake Manitoba carries up to 706 m³ ps of Assiniboine River floodwater.
4 *Lake Manitoba level control* A control structure and channel maintains Lake Manitoba's level between 247.2 and 247.8 m above sea level. Previously, fluctuations were uncontrolled and any over 2 m caused flooding.
5 *Winnipeg dikes* A permanent system of dikes along both sides of the Red and Assiniboine rivers was built to a height 1.22 m *below* the 1950 peak level. City engineers don't expect floodwater to exceed this mark as the Red River Floodway handles enough excess water. Thirty-one pumping stations were also built to return storm and floodwater to the river.

Figure 8.1 Flooding in Winnipeg, Manitoba
In May 1950 (inset), floodwaters inundated most of Winnipeg. During the bigger flood of 1979 (large photo), the spillway of the Red River Floodway discharged much of the floodwater into the channel which bypasses the city. The floodway was so effective that Winnipeg suffered less flooding and damage in 1979 than in 1950.

Figure 8.2 Cross section of the Red River, Winnipeg, Manitoba
Land use of the lower flood plain has been regulated to minimize the damage caused by floods, even those which are as great as the 1826 flow of 6360 m³ps.

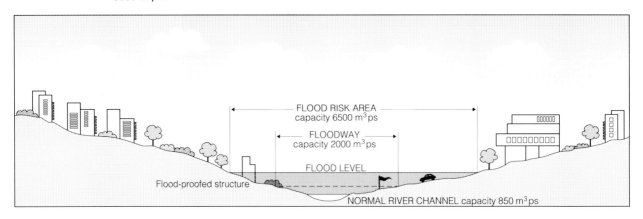

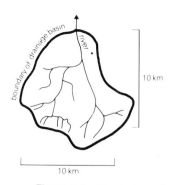

Figure 8.3 River channel area
The area of a river channel is only about 1% of the total area of the drainage basin.

The flood mitigation measures, which cost a total of almost $100 million, won't prevent floods, but they will eliminate the costs and inconveniences of small floods and significantly reduce the damage caused by great floods. The savings from damage by just one great flood may exceed the total cost of these modifications by as much as eight times!

The obstacle course   *The water cycle*

In general, observation shows that a river smooths out variations in precipitation. But first let us be quite clear about the relationship of a river, stream, or creek to the area it drains. The surface area of the water-filled channel is really quite small. In Figure 8.3 a region is shown with only its rivers drawn in. A thick line has been drawn to show the region of land from which water drains to river. We call this area the **drainage basin** of the river. As you can see, the area of water surface is only about 1% of the drainage basin, so although precipitation falls directly into the river channel, by far the largest amount falls onto the land around. Most of this land is covered by soil and vegetation, so the precipitation falling has to overcome a sort of obstacle course to get to the river banks.

The first hurdle   *Infiltration*

Stand out in the rain and the first hurdle of the course becomes clear — you get wet. If you were to stand out in the rain for long enough the rain would soak all of your clothes and then drip off onto the ground, but you would have absorbed a lot of water that will never get to the river. But while you can move out of the rain, plants can't, so they get very wet every time rain falls. And each time, they trap some water before the remainder trickles down leaves and stems and enters the ground (Fig. 8.4).

Figure 8.4 Vegetation surfaces
Water is held as droplets on vegetation. The droplets on these rose leaves will be evaporated when the storm is over.

The giant sponge   *Soil moisture*

The wetting of vegetation is an important delay mechanism and a source of water loss. Nevertheless it is only a minor hurdle in comparison with what is to come. As the water trickles into the soil it enters a world full of narrow tortuous passages, constrictions, and dead ends.

A soil is really a giant inflexible sponge. It has all of the properties of a dish sponge, but you cannot squeeze it to force the water out. Firstly, it soaks up water — pour a little water on at the top and nothing comes out of the bottom. Even when you pour enough water on to cause an outflow there is an inevitable delay as all of the tiny passages are negotiated. Only then does the water come out — the rest is held in the sponge against gravity. Of course it is very important that this should be so, for where else would plants obtain their water?

If you fill a sponge with water and then hang it up it will drain for a long time — a kitchen sponge (10 cm × 10 cm × 0.5 cm) drained for 20 minutes in a classroom experiment. The soil that makes up the drainage basin in Figure 8.3 is 10 km by 10 km by about 2 m thick. A rough calculation shows the soil is about four billion times larger than the sponge. Now perhaps it will be less of a surprise to find that the soil continually drips water into the stream and that a few days without precipitation will not be enough to stop it.

So the soil is able to soak up water, store it, and then let it drip out slowly. This is why rivers can usually continue to flow in periods of drought. However, the Assiniboine River stopped flowing in October 1804 because there had been no significant precipitation for the previous year. And remember that some of the store of water in a soil is sucked out by plants for transpiration, so it does not all drain to the river.

Sometimes, when precipitation falls very heavily, the soil sponge cannot take in the remainder fast enough, so some runs over the surface. In Winnipeg such occasions are all too frequent in late spring when the slow melting of winter snow has saturated the soil sponge. Heavy rains, such as those of May 1974, cause rapid run off of water into the rivers and flooding of all low-lying land. The Winnipeg floodway was created to ease Red River floods by carrying much of the flood water (up to 1700 m^3/second) around the city.

When water runs over the soil surface it moves at about 10 cm a second and this may be ten thousand times faster than through the soil. So when water does flow on the surface it gets to rivers quickly, and floods may result.

Springs ## A soggy start

Water tends to flow straight down a hill slope, using the most direct passages in the soil sponge to get to the river bank. In this way rivers receive most of their water supply through their banks. However, rivers usually start in a hillside hollow: occasionally as a sparkling spring bubbling out of the soil, but much more commonly as a seepage which shows itself as a soggy, waterlogged patch of ground.

Springs and seepages are mostly found in hollows (Fig 8.5). A hollow focusses water to its centre. And, because water from a large region of hillside is gathered together, it is easy to appreciate that the soil in the hollow will not be able to cope with so much water. The surplus water flows out at the soil surface and a new river is born. It is interesting to think that both the Mississippi and the Amazon — two of the world's greatest rivers — start as soggy patches of soil in hollows of thousands of remote hillsides.

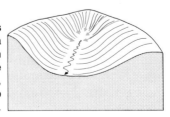

Figure 8.5 Hillside hollows
The arrows show how, in a valley, water moving through soil is concentrated towards one place. When water fills the soil, it seeps out over the surface to begin a stream.

Groundwater

The water beneath your feet

Of all the many pastimes that people enjoy, an unusual one is undertaken by people who put on a tight rubber suit, carry a helmet and a lamp, and then squeeze themselves into dark holes in the ground amid torrents of cold water (see also p. 191). But if these cavers can drag themselves through gaps within the solid rock, what has happened to our soil sponge with its slowly moving water? Well, the sponge is still there, but we shall now examine the rock below.

Most rocks have cracks, but they are too small, too few, and too disconnected to permit the fast movement of water. However, some rocks are different. Limestone has sufficiently large and numerous joints in it for water to seep from the soil deep down into the rock. We shall discover in Chapter 15 how this continual seeping has enlarged the joints to such an extent as to allow whole streams to be swallowed up, but for the moment we should note that water-bearing rocks, called **aquifers,** are quite common. Most of them are made from limestone, chalk, or sandstone (Fig. 8.6). Their main function is to store even more water than the soil could alone. Chalks and sandstones are the most efficient, for they contain countless tiny pores and joints very much like a soil, only the rock may be hundreds of metres thick. Limestones, with their larger channels, pass water more quickly and so do not store it as efficiently.

Aquifer storage has a very important effect on the volume of water in rivers. Because water soaks down into the aquifer and not down slopes to the stream bank, streams on water-bearing rock might receive their water from the rock rather than the soil. This is sometimes very clear when water can be seen bubbling up in the stream bed. Because the rock can store such large quantities of water, it will usually absorb all rainwater. Water is released only slowly from the tiny pores of an aquifer, giving drainage basins with aquifers an unusually good protection against floods.

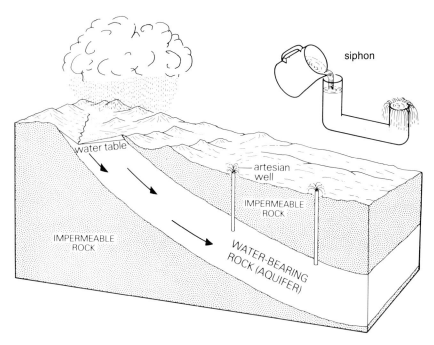

Figure 8.6 Springs The surplus water flows out along a spring line. The remaining water is stored in the rock as artesian water. Wells drilled down to this rock in the nearby lowland will provide water under pressure, much as water poured in at one end of a pipe pours out of the other end of the siphon.

93

Streamflow **Drip, drip, drip**

On its own, a drop of water oozing out of a river bank and into the river would be totally unnoticed. If a drop flowed out of a metre-long bank every second, it would take a whole day just to fill an average-sized bucket. Yet somehow these little drops have to combine to yield all the water that flows in a river. The Red River at Winnipeg, for example, drains a basin which extends west to Alberta's Rockies, east to Ontario, and south to South Dakota. In that area there are over 10 000 km of stream channels and, therefore, over 20 000 km of stream bank. Now if we multiply our single drop in each metre of bank by this figure we find we can collect 2500 buckets of water each second (42 m³/second) at Winnipeg. So it is quite possible for river flow to be provided by slow "insignificant" drops of water oozing out of the banks. Most of the drops flow into the river even though the bank appears dry.

Following a period of precipitation much more water moves through the soil to the river bank. Now water starts to ooze out of the upper part of the bank as well as below river level and the banks become wet. When the number of drops emerging from each metre of the bank reaches the equivalent of about 20 every second, the amount of water reaching the Red River fills the channel. At this stage, it is dealing with 850 m³/second of water, but in the flood of May 1979 at one time there was more than 2995 m³ of water flowing. Winnipeg was saved from serious flooding because the river channel and floodway could handle a total of 3700 m³/second. Not surprisingly, the flat area on either side of many rivers is called the **flood plain.** Fortunately the amount of precipitation is not usually great enough to cause flooding and, on average, streams flood only about once every couple of years.

Flooding is an important natural event, and flood plains are natural temporary reservoirs where water can be held until such times as the river channel can carry it away. Normally this would not matter, but people have a habit of building on flood plains because they are flat. People are not very happy when their living rooms become part of a natural storage system, even for a day or two, so very few river flows are now entirely natural.

Atmosphere and ocean **Connections**

There is a continuous cycle of water involved in the stages described above. Clouds produce precipitation which falls onto the soil and plants. Some is evaporated back to the air and some is used by plants, but most goes on to flow through the soil. This water eventually reaches river banks and causes water to flow in channels. Moving much more quickly now, the river water soon gets back to the sea to form part of the large store of water that is evaporated in order to make most of the clouds. This completes a cycle of water movement (Fig. 8.7).

The water cycle links the big energy flows of the atmosphere with those in the ocean and thus nicely ties up all the loose ends of our present study. Everything is interlinked and is in a long-term balance. However, it is important that we see this as a complete picture. We must realize that if we change one part of the atmosphere-ocean system we are bound to affect others. We need to be sure these effects will not be catastrophic.

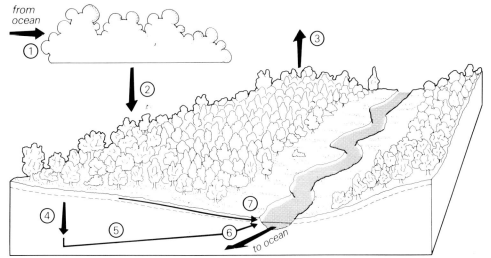

Figure 8.7 The water cycle

SUMMARY: THE WATER CYCLE

Water moves in a never-ending cycle between ocean, atmosphere, and land (Fig. 8.7). As air passes over the ocean it picks up moisture. Air and moisture then rise over the land to give cloud (1). Clouds release the moisture as either rain, hail, sleet, or snow (2). This then falls onto the land surface. Here water may be stored on vegetation and later evaporated back into the air (3). The majority passes into the soil (4). Some is then used by plants and transpired back to the air (5). Where there is an aquifier below the soil, the water passes down to the rock (6). This delivers water to the streams via springs, otherwise it moves in soil to streams (7). Finally it returns to the ocean.

Exercise 8.1   The urban water cycle

1. Study Figure 8.8 and compare it with Figure 8.7.
2. List the differences between the routes taken by raindrops in the two diagrams.
3. Explain why the water in Figure 8.8 will get to the river more quickly.
4. Why should the river in Figure 8.8 be more liable to flooding than that in Figure 8.7?
5. Now study Figure 8.9, which is a photograph of a tributary. Firstly, make sure you can find the river! Now describe the changes that have taken place to the channel of the river. What would you call such a river?
6. Describe the routes water follows after falling on the factory roofs. How much water sinks into the soil?
7. The river is used to carry away surplus rain water. What else has it probably been used for in the past?

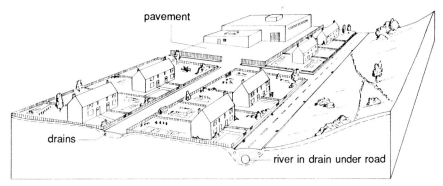

Figure 8.8   The urban water cycle

Using the water cycle

When water becomes our most precious resource

Canadians enjoy easy and unrestricted access to an abundant fresh water supply almost all the time and in almost every part of the nation. And Canadians have become extravagant in the use of this vital resource. For example, during July and August it is not unusual for a domestic household to use in excess of 12 000 L of water each day compared to many Europeans who use only 5 000 L a day.

Severe droughts are uncommon, and usually short-lived in Canada, but an understanding of the nature and frequency of drought should form part of our long-range planning for the most effective use of fresh water. For example, the years 1985-1988 were much drier than normal in western Canada. Many lakes in southern Alberta and Saskatchewan dried up; crops withered and died; fires destroyed thousands of square kilometres of forests throughout the western provinces; many streams shrank until spawning beds dried which killed many eggs and fish and prevented other fish from reaching their spawning beds. In the driest areas, restrictions were imposed by local governments, often on a progressively more limiting scale such as:

1. Hoses and sprinklers to be used only between 7 pm and 7 am.
2. Homes with even-numbered street numbers permitted to water lawns and gardens only on even-numbered days, odd-numbered houses on odd days.
3. Fire hazard warnings issued, and campfires prohibited in forests.
4. The use of set hoses and sprinklers forbidden.
5. Public access to forest areas restricted.

Australians occupy a hot, dry continent and are well aware of the need to conserve water. In times of extreme drought, people are often forbidden to wash cars, water their gardens, and are even discouraged from running taps while brushing their teeth!

Figure 8.9 Water in an industrial landscape at Cornwall, Ontario

Figure 8.10 Movement of water by pipeline and aqueduct

Conflict in water demand

In the years 1975 and 1976 Europe experienced a dreadful water famine. Very drastic measures had to be taken as the ground cracked, flowers died in the gardens, and crops wilted in fields. Industrial and domestic water use had to be rationed in order to conserve water for as long as possible.

The difficulties experienced at this time show what a delicate balance there is between supply and demand. In areas with a regular water supply it is taken for granted that the turn of a tap will produce water, yet this is possible only if rain falls regularly, and even then only because of a vast network of pipes, pumps, and reservoirs that move the water from where it falls to where it is needed.

California uses more than twice as much water as any other American state. So it was in California (where millions of people have settled in an area with little natural fresh water) that the idea of interbasin transfer of water was first developed. In 1913, the Los Angeles Aqueduct began moving 600 million L of water per day from east of the Sierra Nevadas to the city, draining away much of the natural water resources of the Owens Valley. Bringing much greater volumes of water from the Colorado River to the southwestern coast of California was an even bigger project which required the building of 300 km of tunnels and canals.

Co-operation on the Columbia

Water is needed not just for drinking: it is also needed to flush toilets, wash clothes, to cool machinery, as part of many industrial processes, and to irrigate farmland. Rivers are also used for the transport of goods, recreation, and the generation of electricity. So we have to try to manipulate our water resources in order to serve all of these interests. We also have to guard against the occasions when we have too much water, for rivers spill over their banks and flood homes and farmland.

You might think that, for our purposes, the best flow in a river is one that is constant throughout the year. If it were possible to achieve this we would be able to plan for all of our uses to the best advantage. Unfortunately, our demands are not constant. In the Okanagan Valley, for example, more water is needed in summer for the irrigation of tree fruit crops, lawns, gardens, and to cater to the needs of thousands of tourists. Because rain falls irregularly and demands change from area to area and from season to season, we need to store water for release as needed, as well as to prevent flooding. The obvious answer to these problems is a **reservoir.** In the past reservoirs were built only to supply water for drinking. They filled up in winter and the surplus just spilled over the top of the dam. Reservoirs have changed. For example, the great dams and reservoirs of the Columbia River are parts of a major, multi-purpose valley modification scheme. Water conservation, generation of electricity, flood mitigation, and recreation use go hand in hand. Fourteen large dams now control the flow of the Columbia in British Columbia and Washington. When its power stations were more than doubled during the 1970s, the Grand Coulee Dam in Washington State became the world's most powerful hydro-electric generating facility. The expansion of Grand Coulee was made economically feasible by the building of the Columbia Treaty Dams which store spring run-off and moderate the flow of the river. The treaty dams are the Mica, Castlegar (Hugh Keenleyside), and Duncan, north of Kootenay Lake in British Columbia.

Figure 8.11 The Hugh Keenleyside Dam, Columbia River, Castlegar, BC

Figure 8.12 The ultimate in river use
The Colorado River is 2000 km long but it rarely reaches the sea. A series of long lakes store its water for use in irrigating over 1 million ha of the semi-arid lands of southwestern United States and north Mexico and to provide water for drinking and industrial use in the cities of southern California.

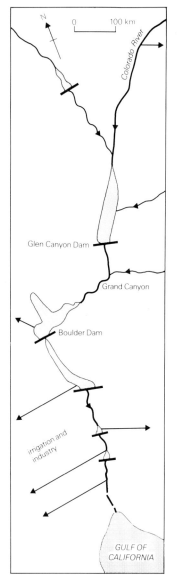

In the USA, the ideas behind multiple use of water were employed in many world-famous schemes, including the Colorado River in Arizona (Fig. 8.12) and a whole series of dams to regulate the Tennessee River. The Tennessee is a large river capable of producing much electricity and it was often deep enough to be used for navigation. The trouble came because the river flow was very variable. There could be long droughts and then a series of violent storms resulting in flooding. So over a 20-year period 31 dams were built to tame the river. Today it looks like a series of long lakes, but the risk of flooding is small and a large amount of power is constantly available to help boost the economy of the region. Locks in mainstream dams aid navigation.

However, there is no point in starting on any regulating program without a detailed knowledge of the way the water cycle works. You could easily find yourself with a reservoir that never fills, or one that floods after every storm!

Exercise 8.2

Figure 8.13 displays the annual course of evapotranspiration (water loss by the soil to the atmosphere), precipitation, soil, moisture use, and soil recharging for Vancouver, British Columbia.

1. During which month does Vancouver experience a surplus of water?
2. When is a water deficit apparent?
3. To what extent does Vancouver have a net surplus or deficit of water?
4. Why is the rate of evaporation less during July and August than April and May?
5. When are Vancouver's soils fully recharged?
6. What happens in Vancouver after the soils are fully recharged?
7. List at least 3 steps that you would expect Vancouverites to have taken to ensure a year-round water supply.

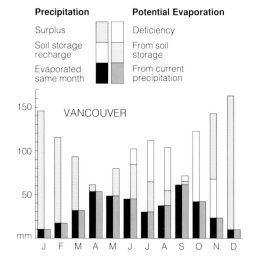

Figure 8.13 Annual water budget for Vancouver, British Columbia

Exercise 8.3

Figure 8.14 displays the annual course of evapotranspiration, precipitation, soil moisture use, and soil recharging for Saskatoon, Saskatchewan.

1. Saskatoon's growing season extends from late April until early October. What are the 3 dominant factors in the water budget during the growing season?
2. Is the water which is stored in the soil during the 5 coldest months all withdrawn during the growing season?
3. Does Saskatoon experience any surplus run-off which could be stored to help overcome the summer deficit?
4. Draw an important implication from your answer to question 3.
5. In a point-form summary, compare the water budgets of Vancouver and Saskatoon.

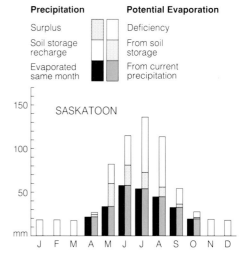

Figure 8.14 Annual water budget for Saskatoon, Saskatchewan

Exercise 8.4 Conserving surface water

Southwestern Alberta experiences a longer growing season, but a more severe water deficit than Saskatoon. Therefore, the irrigation of crops during the summer is an important agricultural activity in the Lethbridge area. Storing spring run-off for summer use is an attractive idea, but there are many costs to be met before benefits can be realized.

In Figure 8.15 two hypothetical proposals for small-scale water storage on the Oldman River are presented. The main reasons for preferring one site over the other are (a) it should not use valuable farmland; (b) it should require only a small dam; and (c) it should have a reliable water supply. State which reservoir site you would choose, and explain the reasons for your choice.

Figure 8.15 Two reservoir designs
(a) Site in the Rocky Mountains; dam height 40 m, length 100 m; land to be flooded 1 km²; reservoir storage 3 000 000 m³, present land use — rough grazing with cattle; number of houses to be submerged — 1.
(b) Site in the foothills; dam height 15 m, length 250 m; land to be flooded 2 km²; reservoir storage 1 500 000 m³; land use — mixed arable and beef farmland; number of houses to be submerged — 10.

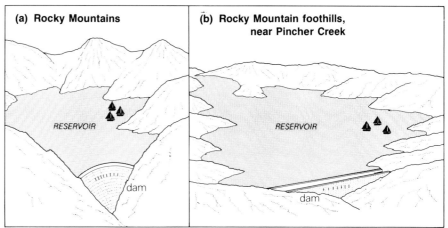

Reservoirs in developing countries

Lessons of Dodoma

Engineers in developed countries, such as Canada and the USA, have a large supply of data to use when designing reservoirs. They have river flow and precipitation data and they also know how much water will be locked up in the soil or used by plants. So, if they do their math correctly, they will end up with a reservoir of the right size for the job. If they understand which rocks are water bearing and which will not let water seep out, then they can also site their reservoir correctly. In developing countries, this information is much harder to obtain and many of the early reservoirs were built by guesswork. Some guesses worked, but the following is the story of one that didn't.

A town in Tanzania (then Tanganyika) called Dodoma needed water. Before 1910 people depended on an irregular supply from rain ponds which dried out in the dry season and after that they had to dig holes in river beds. Around 1910 Dodoma became an administrative centre and the population increased, so a better water supply was needed urgently.

A reservoir was begun in 1929. Unfortunately at this time there were no long-term records of precipitation or evaporation, so the size of the reservoir had to be estimated. Eventually a guess was made and a dam constructed to hold water only to a depth of 12 m.

The region receives about 500 mm of rain a year. This is the same as the Cariboo-Chilcotin area in BC, but Tanzania is a very hot country and evaporation is high. Also much of the rain comes in heavy downpours over just a few months. This means that, while rain is falling, there is a lot of water that runs straight over the surface. The result is that relatively little sinks into the soil to keep streams flowing in the rainless months. So, the rain reaching the reservoir in the rainy season has to last all year and this can be quite a problem with such a high rate of evaporation. In addition, the drainage basin was forested, so much of the water stored in the soil was taken up by tree roots during the dry season.

As a result of all of these factors there was insufficient water to fill the reservoir in years with low precipitation, and constant use by the town caused it to run dry. In an attempt to increase the water reaching the reservoir the trees were cut down. More water did, indeed, subsequently reach the reservoir, but it was full of mud eroded from the now-bare soils. So in a short time the reservoir began to silt up badly and its storage capacity was severely reduced (Fig. 8.16). By 1944, a second reservoir had to be built on a new river. Using lessons derived from mistakes with the first reservoir, the new one was arranged to drain a much larger area, so the reservoir filled even in dry years, without the need to cut down the trees.

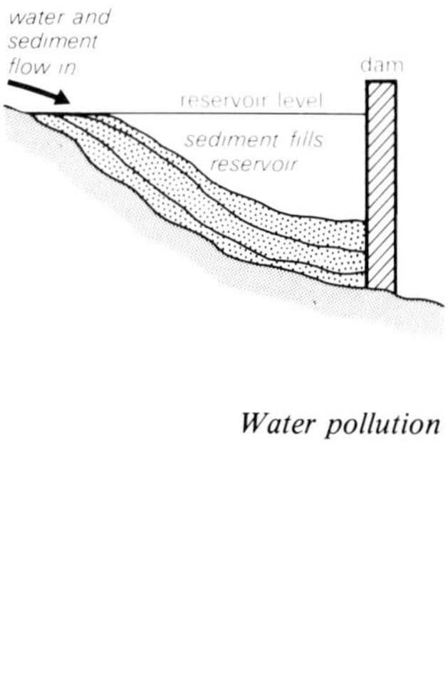

Water supply can be effectively designed only with a full knowledge of how the water cycle works. Unfortunately, the data needed for accurate design are still not available in many developing countries, and without them improvement schemes can never be fully effective.

Figure 8.16 The effect of sediment accumulating in a small reservoir
Large reservoirs also silt up, but more slowly.

Water pollution

Rivers — R.I.P.?

In 1775 the English inventor Alexander Cumming patented a flush mechanism for toilets. This was improved a century later by Thomas Crapper. Unfortunately, these developments hastened the decline of many of our rivers into sewers.

In the 19th century the rivers of the developed world provided power to turn the wheels of industry, carried its goods, provided raw material for the chemical

Figure 8.17 The river as a means of disposing of wastes

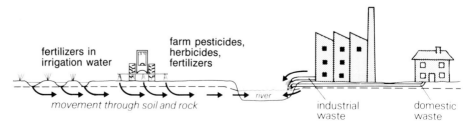

industry, and were a convenient means of disposing of its waste (called **effluents**) (Fig. 8.17). But toxic waste from factories killed off the bacteria that decompose dead organic matter naturally and safely. So in the late 1800s, when people built indoor plumbing systems and flushed human waste into the rivers, there were not enough natural organisms to decompose it. And so the stench rose, disease was spread and the fish died. The rivers had been killed because of greed and a lack of understanding and until the middle of this century very little was done about it. By 1958 even vast expanses of water like the 25 000 km² Lake Erie were dead around the edge and dying in the middle. And, as if to symbolize the disaster of river pollution, in 1969 a river in the USA covered with oil and floating debris caught fire!

Of course pollution was worse in the summer when river flows are low, and worse still at the tidal limit where sewage floated out on the ebb tide and back in on the flood tide. There was a vital need to understand how the water cycle worked and what limits could be imposed on it.

It was not just industry and households that were to blame. For over half a century we have been pouring increasing quantities of fertilizer, pesticide, and herbicide onto the land. This is washed through the soils by precipitation and, of course, it eventually finds its way to rivers. The fertilizers provide food for algae which then multiply and turn the water green. They use up all the oxygen in the water so there is none left to help in decomposing dead organic matter properly, and sewage just floats in the water. So rivers have been killed by agricultural pollutants in rural areas, industrial pollutants in citites, and human sewage in both.

You can see the geography of pollution by looking at any atlas map. In Europe, North America, and any other place with industrial development beside rivers, you can almost guarantee that pollution will occur. In the Netherlands the Dutch have stopped taking water from the Rhine: it is so polluted by German and French industry that it has become too costly to filter, clean, and sterilize for use as drinking water. Britain and Canada now lead in trying to cope with this problem, and the Thames River, for example, has been cleaned up to allow fish to thrive once again even in the City of London itself.

Improvement in water quality has mostly been achieved through laws controlling the pollutants that industry may discharge into a river and by building better sewage works. Our rivers are very forgiving, but they can only cope with a certain load of waste, and no more. Now we understand what these limits are, we are in a better position to allow the rivers to return once again to their natural condition and be of benefit to us at the same time.

Irrigation problems

Water — handle with care!

Throughout the world, more water is used for irrigation than for any other purpose. Just six countries (China, India, USA, Pakistan, USSR, and Iran) account for over 70% of irrigated land, 37% (84 million ha) in China alone. Here, more than anywhere, water is life. Traditional irrigation methods allow water to flow into a field which has a low bank built around it. The water ponds up and sinks slowly into the soil (Fig. 8.18). Very often water is led

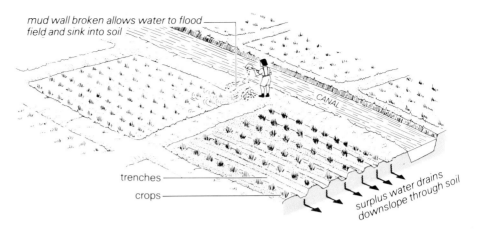

The principle of irrigation in developing countries.

Figure 8.18 Simple irrigation methods

Traditional irrigation uses employ animal power to lift water from a river and pour it into a canal. Fields are irrigated by digging holes in the channel banks. This photograph was taken in India.

along trenches or furrows dug between growing crops. All irrigation systems need a lot of attention. Usually water is fed into a field by digging a hole in the bank of a nearby canal and the flow is stopped by filling up the hole. In all cases water on the surface evaporates rapidly, so much more water needs to be applied to the land than the plants really need. And by diverting all this water from rivers onto the fields, little is available for users downstream.

Some of the world's irrigation schemes are on a massive scale and they represent one of our greatest influences on the water cycle. In the Indus valley in Pakistan over 100 years of effort has provided water to irrigate more than 10 million ha of land (Fig. 8.19).

However, irrigation can also bring problems. With large schemes so much water is poured into the soil that it may cause a permanent rise in soil water levels (Fig. 8.20). In downstream regions water can even be forced to the surface. But this water contains many salts that have been washed out of the rocks and are now brought to the surface in such quantities that salt crusts form on top of the soils and plants will not grow.

Great care must be taken with irrigation to avoid this problem, otherwise as new areas are developed for irrigation, others will be destroyed and abandoned.

River control

Taming the Nile

The demands for irrigation can sometimes be so great in arid and semi-arid areas that the water cycle is disrupted completely. In the southwestern United States, the famous Colorado River rarely reaches the sea because all of its waters are usually diverted along canals to irrigate California and Mexico (Fig. 8.12). Even the African Nile, one of the world's largest rivers, does not reach the sea for three months each year.

The Nile begins its course in the mountains of Central Africa (Fig. 8.21). Here, although precipitation is plentiful, it is seasonal, providing a summer flood to the Nile but giving much less water in winter. As the Nile flows

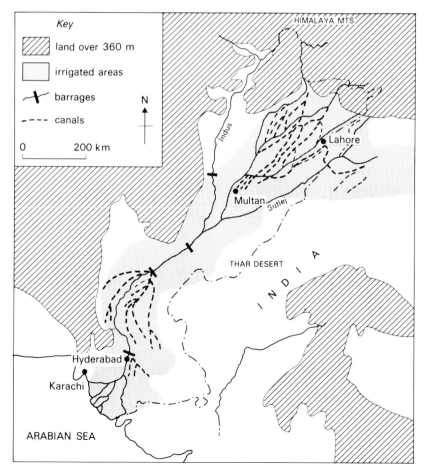

Figure 8.19 Irrigation of the Indus valley
Here 60 000 km of canal irrigates more than 10 million ha of land.

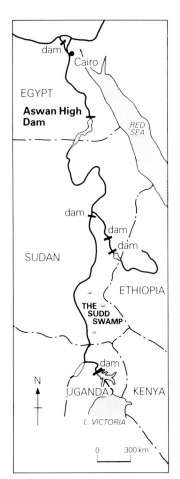

Figure 8.21 The Nile and its international tributaries

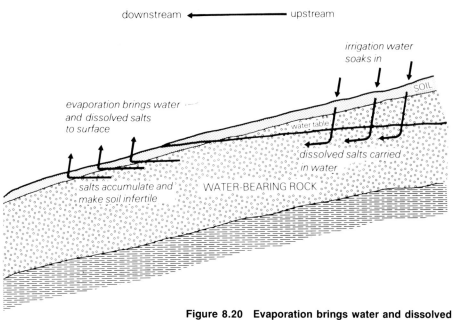

Figure 8.20 Evaporation brings water and dissolved salts to the surface

north towards the Mediterranean Sea, the last 2000 km of its course are through desert. Here the river provides the only means of life to the millions of people who live near its banks. Irrigation of the lower Nile valley, and especially the Nile delta, goes back over 3000 years.

Today the population in Egypt is over 35 million and still rising, and the need for food is getting ever greater. Traditionally farmers would irrigate fields whenever there was enough water in the river, but the variable supply did not help agriculture to be as efficient as it might. This provided the Egyptian government with a dilemma: how could they provide more water when the natural flow of the Nile was already fully used? The answer seemed to lie in building a big dam to trap the flood waters and to use this "surplus" to irrigate new lands and allow more regular supply to those areas already being irrigated.

In 1953 the Aswan High Dam was built and the Nile was tamed. Now new lands can be irrigated. But there has been a cost to pay. The reservoir (called Lake Nasser) is filling up with sediments that used to make the soil of the valley fertile. The farmers no longer get free fertilizer. They have to buy expensive artificial substitutes which also cause environmental damage.

Sudan has also built dams on the Nile to irrigate its northern lands. South of the dry lands that need irrigation there is a huge swamp called the Sudd. Now a project is under way to dam part of the Sudd for agriculture and release some of the swamp water for irrigation downstream. To do this the Jonglei Canal is being dug right through the swamp. But again, benefits to some bring hardship to others. The local Dinka tribe will have to stop herding cattle near the swamp and, instead settle and cultivate crops. The use of yet more water for irrigation in these upstream regions leaves less for downstream areas.

QUESTIONS

Multiple Choice

1. Even when clouds are not present, water vapour and carbon dioxide in the air reduce the rate of heat loss creating a phenomenon called
   (a) albedo
   (b) greenhouse effect
   (c) radiation
   (d) condensation
2. The uneven heating of the earth sets winds and ocean currents in endless motion to achieve
   (a) Coriolis effect
   (b) chaos in the mid-latitude gears
   (c) latitudinal transfer of heat
   (d) condensation along coastlines
3. Fog occurs where cooling air
   (a) drops its temperature below dew point
   (b) mixes with warmer, rising air
   (c) passes over a warm ocean
   (d) rises along a cold front
4. Tropical cyclones tend to travel
   (a) towards the equator
   (b) towards the northwest
   (c) west, polewards, then east
   (d) east, equatorwards, then west
5. India receives monsoon rains when
   (a) the jet stream reverses direction
   (b) the jet stream crosses the Himalayas
   (c) the jet stream breaks down and disappears
   (d) the Himalayas are at their coldest in May
6. Most of the world's ocean coastal areas experience two high tides and two low tides each day because the earth's seas and oceans are shaped into two bulges and two hollows by
   (a) gravitational effects of sun and moon
   (b) centrifugal forces caused by rotation
   (c) Coriolis Effect which differs in each hemisphere
   (d) uneven heating of the ocean surface
7. The distribution of water pollution is closely related to
   (a) industrial development in Third World countries
   (b) industrial and urban activities along the stream banks
   (c) the careless use of water for irrigation
   (d) the circulation of great ocean currents
8. The damming of the Colorado River and the use of progressively greater amounts of its water for many different activities have caused
   (a) increased agricultural and urban activity in southern California
   (b) a warming of the climate, and the desertification of Mexico
   (c) the Colorado to flow to the sea for only 6 months a year
   (d) salts to diminish in the Salton Sea

Short Answer

1. Explain what is meant by (a) a depression and (b) an anticyclone.
2. Draw a diagram of a complex frontal system in a depression. Show the main types of cloud that are produced at each front.
3. Explain why rain is produced at frontal regions.
4. Why do anticyclones usually create fine, warm weather in summer but cold, clear weather, often with fog, in winter?
5. Explain why Los Angeles, located in a coastal basin, is particularly prone to atmospheric pollution.
6. Name one area which frequently experiences tropical storms. Draw a diagram to show how such a storm forms, and a map to show its typical path.
7. Explain what is meant by a maritime climate and a continental climate. What are the main differences you would expect to find between the climates of Vancouver Island and southern Alberta?
8. Name two countries that have monsoon climates. Draw a sketch map to show their locations and seasonal winds.
9. Explain the main differences between an equatorial and a monsoon climate.
10. Draw and label 3 hot deserts on a sketch map. Explain how sinking air is responsible for their formation.
11. Name a dam and reservoir that is used for several purposes. Explain how it is operated.
12. In diagram form, explain how artesian water is obtained.
13. Describe a primitive form of irrigation which is still in use today. What is one disadvantage of this system?
14. Why is it difficult to rationalize the use of dams for irrigation, hydro-electrical generation, and flood mitigation?
15. Why is salinization a major hazard in many irrigation areas?
16. Draw and label a diagram to explain the causes and effects of rain shadows in the Okanagan Valley and southern Alberta.
17. State a generalization about the variations in temperature with changes in each of the following factors:
    (a) altitude; (b) latitude; (c) aspect; (d) atmospheric transparency.

Part three: Soils, vegetation, and people

All things form delicate links between the earth and the atmosphere. Each species has slowly adapted itself to occupy a special place within the world's landscapes.

Figure 9.0 Contrasts in ecosystems

9 Weathering

Weathering

A grave lesson

When archaeologists dug down under the remains of an Iron Age fort in southern England they found much evidence of past occupation. There were bits of broken pottery and a few coins from a long-forgotten civilization. But most interesting of all was the skull and bones of a man aged about 25 who had been killed by a blow on the head with an axe.

These gruesome finds have important lessons for the geographer as well as the archaeologist. Notice what was found — bones, coins, and pottery. Presumably when the body was buried it was complete with flesh and clothes. Since the burial, the skin and clothes had rotted away and the coins had lost much of their surface markings. By contrast, the pottery was still intact and had not even lost its decoration. Clearly, organic matter — flesh and clothing — very readily decomposes (rots away) whereas bones, which are mostly made from inorganic materials, such as calcium carbonate and calcium phosphate, decompose more slowly. Nevertheless the bones were showing signs of considerable surface attack. The coins had started to lose their surface marking, indicating that metal is attacked by chemicals in the soil. The clay pottery was untouched, so clay must be more resistant than all the other materials.

Natural breakdown processes are divided into **physical weathering** (the *disintegration* of rock into smaller fragments) and **chemical weathering** (the *decomposition* of rock or organic material into new materials). In the grave we noticed only evidence for chemical decay, showing that soil cover prevents physical weathering. Physical weathering is important only in places where bare rock can be found, such as on a mountainside, a cliff, or in a desert.

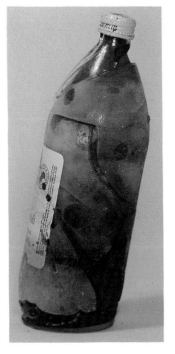

Figure 9.1 Frost shatter

Frost shatter

A chip off the old block

Many parts of the world experience cold, frosty conditions at night. A frosty night will sometimes cause ice to form on wet roads and be a danger to drivers. It will also cause rain water to freeze in the cracks of a rock.

When water changes to ice it expands by about 10%, creating an extremely powerful and destructive force. Figure 9.1 shows the reason you are told not to put full bottles of liquid in the freezing compartment of a fridge. As water freezes it expands and cracks the bottle. The process is called **frost shattering**.

A bottle is simply a piece of transparent rock, because glass is mostly made from sand. As a result we can compare freezing water in the confined space of a bottle with water freezing in the cracks of mountainside rock. In nature all that is needed, therefore, is rock with some cracks in it, water to fill them, and a large natural freezer. One of the best places to find such a combination is in a mountain area in autumn or spring, when day temperatures can be above freezing but nights are usually still frosty. Any rain or snow melt will seep into the cracks and be frozen at night. As the water freezes it expands and it may wedge pieces of surface rock

Figure 9.2 Frost shatter
A block split by freeze/thaw. The black camera case gives the scale.

apart (Fig. 9.2). This provides room for more water the next time rain falls or snow melts. During the day the ice melts and water may also be able to penetrate farther into the rock, before freezing again at night. After many freeze/thaw movements, pieces of rock may be dislodged so far that they become unstable (Fig. 9.3). The next time the ice melts and releases these pieces from their frozen prison, gravity will cause them to fall away, at the same time exposing new material for shattering.

Frost shatter is most active on mountains and at high latitudes where the temperatures hover around freezing for several weeks at a time. However, the process of frost shatter will affect any exposed rock. Many quarry faces and coastal cliffs have small piles of shattered rock at their bases after a period of frost. These piles of rock fragments are called **scree**. Frost shatter even affects some roads in a very cold winter. If the water in the soil beneath a road freezes, it will expand and lift up the surface, usually causing the asphalt to crack. In Canada, Scandinavia, and other northern areas the repair bills for frost-shattered roads is a large part of the total road costs.

Exfoliation

Rock flakes

Much of the income from Middle-East oil is being used for building factories, offices, and homes in what were underdeveloped countries only 40 years ago. The only trouble is that some of the buildings are breaking up near

Figure 9.3 Frost shatter
(1) This rock has joints which allow water to seep in. During the winter the water expands as it turns into ice. (2) As a result some blocks are forced away from the main mass of rock. (3) Eventually they fall down to accumulate as scree at the base of the slope.

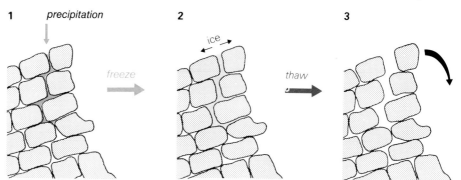

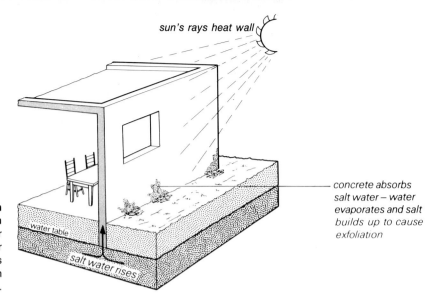

Figure 9.4 Exfoliation
Water containing salts has been seeping up these walls for several years. As the water evaporates, the salt crystals force the wall to break off in sheets.

their foundations. Great flakes of concrete are being shed from new buildings (Fig. 9.4), causing the people great concern. The process at work in these "flaking buildings" is a speeded up demonstration of a common desert phenomenon called **exfoliation.** It happens whenever water seeps into rocks that are heated and cooled rapidly.

Deserts experience great heating by day and cooling at night. As a result, rocks continually expand and contract. When the rocks are heated they swell, pores and cracks open, and it is easier for rain water to get inside. Although rain is rare in deserts and does not last long, some water will always soak into the rock. But rain water is often a mixture of water and sea salt. Soon the water is evaporated, but the salt is left inside the rock (Fig. 9.5). When the rock cools again at night it cannot contract properly because of the salt, and great stresses are set up within the surface layers. The next time rain falls, the process is repeated and more salt is deposited in the rock. Eventually the rock cannot withstand the forces any longer and it fractures parallel to the surface, producing flakes usually a few centimetres thick (Figs. 9.5 & 9.6). Surprisingly perhaps, heating and cooling without rain in a desert does not cause exfoliation. As a result, it is easier to see the effects of this type of physical weathering near desert coasts where there is more salt and more moisture, or near oases and other places where water comes close to the surface. The hottest and driest parts of deserts probably produce very little weathered rock.

Figure 9.5 Exfoliation
With repeated showers of rain salt crystals grow and help wedge scales of rock from the main mass.

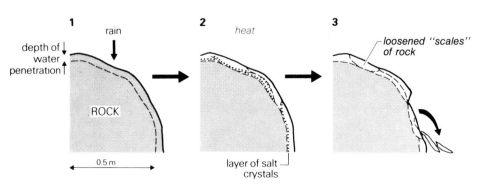

Figure 9.6 Exfoliation of a massive sandstone
The surface of Ayer's Rock, Central Australia, shows well developed exfoliation. Hikers and climbers must be careful on such loose, flaking surfaces.

Chemical weathering

The rot sets in

There is much to be thankful for in the slow rate of physical weathering in a desert. Because of it the pyramids and temples of Egypt still look much as they did 3000 years ago; even their painted surfaces are sometimes fresh and brilliant. In the middle of the 19th century, one of the Egyptian monuments was moved to New York City and erected in Central Park; the people of New York nicknamed it "Cleopatra's Needle." Within a few years it had become so weathered by chemical rotting that the surface writing could hardly be read and restoration work was necessary. (See Fig. 9.7 for another example of chemical weathering.)

Figure 9.7 The result of chemical weathering over the past 100 years.

We all know of extreme forms of chemical weathering. In industry, acids are used to "clean" metals and etch glass. The most obvious example in our lives is the speed with which cars and trucks rust when road salt is used during the winter. In nature there are many less rapid but nonetheless effective forms of chemical weathering. The most important is a reaction between rock and the very weak acid that we call soil water.

If a piece of rock is kept in a tank of rain water there will be a very slow chemical reaction between the water and the rock surface. After a while — perhaps a few months — although you would probably not see a change, a chemical analysis of the water would show that it contained more soluble materials than at the start of the experiment. These soluble materials were once contained within the structure of the rock and have formed because of chemical weathering. A microscopic examination of the rock would show the surface to be pitted. The pits show where the rock has been weathered. If rain water was left in contact with the rock for 100 years, a much more obvious change would probably occur. The surface of the rock would be covered in a loose, very fine material called **clay**. Clay and

Topic   Rocks, blocks, and decay

Weathering is a surface reaction, so the greater the surface area of rock, the faster it weathers. Rocks are formed deep below the earth's surface under great pressure. Often the pressure causes sedimentary rocks to fracture along lines of weakness. A rock made from many bands of sediment will have many weaknesses and will fracture into small blocks (Fig. 9.8). A rock made from a uniform material will have fewer cracks in it. Igneous rocks were once molten and their fractures come from expansion as batholiths are exposed by erosion (Fig. 9.9).

When rocks come to the surface, there is much less pressure on them, and the fracture lines open up, providing places for chemical weathering to work. Rocks that have many fractures (**joints** and **bedding planes**) will weather rapidly. Granite has few joints and this helps slow down the rate of weathering.

Figure 9.8   Soft sediments in Death Valley
Joints and bedding planes are apparent in these rapidly weathering shales, limestones, and sandstones near Furnace Creek Inn, Death Valley, USA.

Figure 9.9   Granite weathering
Exposed batholiths undergo massive expansion and exfoliation. This cracked granite is part of the Sierra Nevada.

solutes are the two main products of chemical weathering.

Although pure water will react with rock, usually rain water can be regarded as a dilute acid, because it has gases such as carbon dioxide from the air dissolved in it. Carbon dioxide gas combined with water produces carbonic acid, although it is extremely dilute. Near industrial areas or volcanoes there will also be sulphur dioxide in the air, and when this is chemically combined with water it ultimately makes sulphuric acid. These dilute acids increase the rate of chemical weathering. As you would expect, the stronger the acid, the faster the weathering. Indeed, it has been the high level of pollution from sulphur dioxide that has speeded the decay of "Cleopatra's Needle" and many other monuments and buildings.

There is a further source of acidity. On the surface of the soil there is often a layer of partly rotted leaves; this is what gives the "springy" feel if you walk through a woodland. As these leaves rot they release acids which combine with the rain water as it soaks through to the soil. They act in the same way as the gases combined with water but they are much more powerful. Thus, water moving through a soil is armed with two types of acid. These are the acids that can weather rock and produce new soil.

Chemical weathering affects the *surface* of a rock. Rocks that are well jointed therefore provide many more surfaces to attack than does a rock with few joints. Although all rocks are subject to chemical weathering, its effect on joints can be seen particularly clearly in some limestone areas (Fig. 15.1). The soil and vegetation have been stripped away by glacial erosion and the rock surface exposed to view. Chemical weathering has enlarged every joint, making the area really quite difficult to walk over.

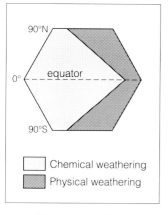

Figure 9.10 Weathering
This simplified model of the world illustrates the relationship of weathering and latitude.

SUMMARY: WEATHERING

Weathering is the natural breakdown of rock. *No movement is involved* and, unless material is transported away, weathered material will just build up where it is formed. To speak of a rock being eroded it must be both weathered and transported from the place of weathering.

Physical weathering is the disintegration of rock by mechanical force. This can be the result of water freezing and expanding in cracks (Fig. 9.3) — a process called **frost shattering** — or of the deposition of salt in rocks that are expanding and contracting in a desert — a process called **exfoliation** (Figs. 9.5 & 9.6).

Physical weathering can produce only fairly large fragments. Anything smaller than a grain of sand is usually the result of **chemical weathering**. Chemical weathering is the chemical reaction of water or a weak acid with rock. This reaction forms totally new materials. Rain water is normally a weak acid because it has carbon dioxide gas combined with it, making carbonic acid. Water passing over decaying vegetation also produces acids. Both types of acid can be very powerful in rotting rocks, even at depths of many metres. The new products of chemical weathering are mostly very small particles called **clays**. The material left over after the clays have been formed is usually soluble and is carried away in solution with the rain water.

In places with a vegetation cover, chemical weathering is more important than physical weathering. Soils are largely the result of chemical weathering.

10 Soil formation

Soil components

A little bit of this and...

As weathering occurs all over the earth's surface, its results should be fairly easy to see and have a well-known name, such as the scree mentioned in the previous chapter. Scree slopes form when many pieces of physically weathered rock have accumulated at the base of a slope (see page 158).

Scree slopes are common in mountain areas and at the base of some sea cliffs — but it is unlikely that you pass a scree slope on your way to school! The absence of screes from most areas tells us that physical weathering is *not* the most usual form of weathering. The other form is chemical weathering, but if this is so widespread, where can we see it? The answer to this is "more or less everywhere," because the usual term for chemically weathered rock is **soil**. This is an important and often overlooked fact, so the next time you go into a garden remember that you are looking at plants growing in rotted rock.

Soil is the normal product of chemical weathering, but as it forms it covers up the solid rock, preventing physical weathering processes from operating. So the rule is: *where you find bare rock, physical processes dominate; where the rock is buried beneath soil, chemical weathering processes have taken over.*

Soil consists of a whole range of "bits and pieces." There are some large pieces (which we commonly call **sand, gravel,** or **pebbles**) that are made of solid rock which has not yet been completely weathered; there are the very small **clay** pieces that are one of the products of chemical weathering, and there is quite a lot of water (which is actually performing the weathering) in which the soluble products of weathering are found (Fig. 10.1). Then, besides all of these inorganic pieces, there are lots of tiny fragments of dead plant together with earthworms, woodlice, mites, and many tiny organisms. If we use a microscope to look at a piece of soil, we see it crowded with tiny living things. There will probably be millions of white threads of fungus, tens of millions of bacteria, and many other organisms (Fig. 10.2).

It is worth taking a closer look at soil. Imagine a world where you have been shrunk one million times. You will then be the size of a large clay

Figure 10.3 (Right) Leaf decay
(1) As soon as the leaf begins to die it shrivels, dries up, and falls to the ground. (2) The leaf is then decomposed by soil micro-organisms and gradually loses its identity. (3) The decomposed fragments (called humus) are taken into the soil by earthworms, digested, coated around other soil particles, and thoroughly mixed with the rest of the surface soil, (4).

Figure 10.1 The components of a soil

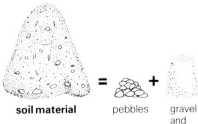

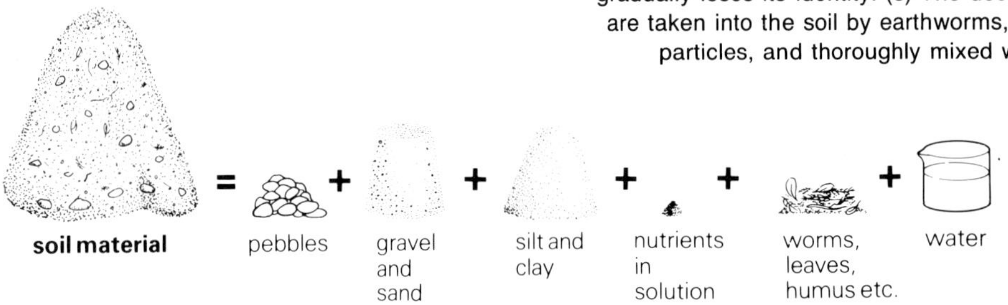

soil material = pebbles + gravel and sand + silt and clay + nutrients in solution + worms, leaves, humus etc. + water

particle! The soil suddenly becomes a world of massive boulders (sand grains), with an intricate network of caverns between. We would call these caverns **soil pores,** and they are important because they allow rain water to soak down into the soil while allowing the surplus to drain right through. If you were one of the millions of small soil animals you would regard these pores as your home and also your hunting ground for food. If you were a plant root, then they would be easy paths along which to grow as well as sources of water and air.

Both plants and animals need oxygen in the soil pores to survive. Most plants cannot survive complete waterlogging of a soil for long periods because they cannot get enough oxygen to their roots. There are some exceptions, such as rice and the plants that grow by the waterside of most streams and ponds, but a plant needs to be rather well adapted to survive in a very wet place.

From a microscopic viewpoint the soil's inorganic bits and pieces provide a sort of hotel for an amazing variety of small animals and also plant roots. But both plants and animals need to live, so where is the source of food?

Figure 10.2 Soil animals
This strange object looks as though it might have been invented for a science fiction film. It is, however, an extremely common soil organism (a protozoan) enlarged 12 000 times. There are myriads of such beautiful and mysterious objects in every gram of soil. The twisted threads are pieces of fungus.

Hotel Animalia   *Soil life and organic matter*

The soil contains a whole army of small creatures which decompose dead plant material and turn it into a form suitable for plant roots to absorb. Some of the creatures, such as earthworms, are easy to see, but most can be seen only under a magnifying glass or a microscope. A magnifying glass focussed on a piece of soil will probably show little orange blobs of "jelly": these are mites. There will also be little bits of "animated hair" which are really nematode worms. But these are only the beginning. Every cubic metre of soil contains tens of millions of such creatures, all workers in the soil hotel.

The soil hotel contains various departments. On the surface is the storeroom of dead plant matter such as dead leaves, twigs, and tree trunks. There is a continual supply of material to this surface store, but it is quickly removed (Fig. 10.3). Earthworms, ants, mites, and other creatures drag plant tissue underground into burrows; if it is too large, they chop it up on the surface. These creatures are like waiters carrying food from the storeroom to the kitchen. Their burrows also help keep the soil ventilated and drained. In the burrows lie millions of bacteria, fungi, and other micro-organisms. These are the kitchen staff. They decompose the organic material and produce a sticky substance **(humus)** which stains the soil black. They also produce a range of soluble foods called **nutrients.**

The guests in the hotel are the plant roots. They can take in only soluble foods, so they suck up water which contains nutrients released by decomposition. The plant roots spread out in a very efficient net, delving into a multitude of soil pores especially to find more soluble food. The nutrient foods they find are sent up the plant roots and used to help plant growth. However, eventually these plants also die and their remains are added to the surface storeroom of the soil hotel for future use by other guests.

Completely decomposed organic matter is called **humus.** It performs three vital roles in a soil: first it sticks soil particles together; secondly, it provides

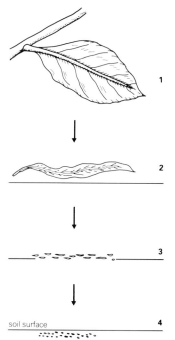

Figure 10.4 Soil binding
When you pull up a clump of grass, notice how the fine roots intertwine and hold the soil. Notice too how the soil has formed itself into clumps, so making the binding task of roots much easier.

an additional supply of nutrients for plant use; and thirdly, it helps to retain moisture. Humus is easy to see in a soil because it is black. When mixed with iron compounds (which are a dull-orange colour) the soil becomes dark brown. Earthworms are mainly responsible for mixing humus, iron, and rock fragments. If you look at the sides of a pit dug into a soil, you will see that the surface layer (which has had the "earthworm mixing treatment") is dark brown, whereas at depth it is much lighter (Figs. 10.7 & 10.9). Earthworms don't like living at depth because their "restaurant" is on the soil surface. The lighter brown colour of the lower layer is the true colour of the inorganic fragments without humus staining.

The sticky nature of the humus is vital because it prevents individual soil particles being washed or blown away. Also "clumps" of soil particles do not fit together very well and they provide gaps through which water can drain. Next time you are in a garden, pick up a small piece of soil and look at it carefully. You will see that it is made from a range of different particles, all invisibly gummed together. You can rub the particles apart, but the gum is sufficient to keep the particles stuck together in clumps within the soil (Fig. 10.4).

One thing after another *Soil layers*

As we have seen, soils begin from the weathered products of rock. Sometimes the rock is not solid, such as when it has been deposited in fragments by a glacier as **moraine** or **till,** by a river as **alluvium** or by wind as **loess.** Nevertheless, in all cases the material from which a soil forms is unorganized and it doesn't contain organic matter. It is called the soil **parent material.**

Soils are divided into three types of layers: the surface layer containing a mixture of plant and weathered rock material; a middle layer beneath the surface which contains clay and partly weathered rock; and a bottom layer — the parent material — which has not been weathered at all. We use letters as a form of shorthand for these layers: **A** for the *surface* layer; **B** for the *middle* layer; and **C** for the *parent material* (Fig. 10.5).

Soils form slowly. Physical and chemical weathering may have to break down the parent material if it is solid rock. In any case the soil begins life as only a C layer (Fig. 10.5a). Gradually, perhaps over hundreds of years, a weathered surface layer is formed and this provides the nutrients that plants need in order to live. Soil organisms begin to invade the soil to decompose any dead plant tissue and mix it with the weathered surface rock.

Figure 10.5 Soil development
Soil gradually develops from (a) unweathered rock to a fully developed zonal soil (c).

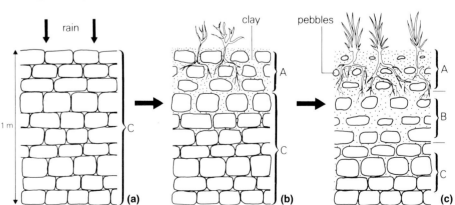

Figure 10.6 A brown earth soil
The (a) mineral and (b) organic parts of a soil combine to make up (c) a complete soil. In diagram (c) the soil layers are a mixture of mineral and organic material (A); weathered mineral matter but very little humus (B); solid rock, yet to be weathered (C).

This mixed layer of plant and rock material is the A layer and it forms above the C layer (Fig. 10.5b).

As more time passes so the depth of weathering increases, helped by acids produced from decaying plants (Fig. 10.5c). Because plants need light and their roots need air, roots are mostly confined to the surface layer. The deeper part of the weathered rock therefore contains little organic matter and is called the B layer. Slowly the B layer becomes thicker. Most soils contain well developed A, B, and C layers (Figs. 10.5c & 10.6c). However, there are many variations to the thickness and properties of these basic layers, as we shall see.

Down in the valley   *Lowland soils*

Many lowland soils in Canada are called brown earths because brown colours dominate the soil profile. They typically have A, B, and C layers as shown in Figures 10.6c and 10.7. Brown earth soils develop on parent materials that weather to release a large supply of nutrients, making them valuable soils for agriculture. Many of these soils are also deep, providing good anchorage for plant roots and a store of water for plant use between periods of rain.

Figure 10.7 A brown earth soil from a lowland area
Compare this with Figure 10.6c.

Exercise 10.1   When is a soil not a soil?

You will need to read the sections "Down in the valley" and "In upland areas" before completing this exercise.

Soil is a mixture of plant and rock fragments, most of which have been decomposed in some way. Soils are most easily distinguished from one another and from other materials by digging pits and looking at the changes that become visible at depth.

1. Study the three sketches of materials found after pits had been dug down to a metre below the surface (Fig. 10.10). State which one of them is a soil, a scree, and a peat. Explain your choice.
2. Study Figure 10.6a. This shows the distribution of rock fragments that would be expected in a soil from lowland areas.
   (a) What are the main sources of acids used in rock weathering?
   (b) Why should there be few pebbles (unweathered rock fragments) near the surface?
   (c) Why does the proportion of unweathered rock increase with depth?
3. The sketch in Figure 10.6b shows the distribution of organic matter.
   (a) Why are most of the decomposed plant remains near the surface?
   (b) What do you think the few large roots belong to? Why should they be found at greater depth than others?
   (c) Why are worms and microscopic soil animals concentrated near the surface?
   (d) What do we call decomposed plant tissue?
   (e) What effect does decomposed plant tissue have on soil colour?

Figure 10.8 A tropical red soil, Malaysia
The trees on top of the cliff are 30 m high and there are steps cut up to the house. The rock of the layer is a light-coloured granite.

Lowland soils in tropical areas such as Malaysia and Nigeria also have soils with A, B, and C layers (Fig. 10.8). The main difference is that chemical weathering under tropical conditions releases red-coloured iron oxides into the soil instead of the brown or dull orange that occurs in Canada. Because of their colour these soils are called **tropical red earths.** (These soils used to be called laterites.) Another difference is that soils in mid to high latitude countries may only be 1 m deep, whereas those in the tropics may be many times this value.

In lowland areas with much less precipitation, such as the American Great Plains or the Russian Steppes, soils are not brown or red but black at the surface. The Russian word **chernozem** is used for these black soils (Fig. 10.9). The black layer may be 1 m deep and it is mostly coloured by humus. These deep A layers develop because of the way soil animals are influenced by climate. Summers are scorching hot and quite dry, whereas winters are very cold and the upper part of the soil is frozen solid. By contrast, spring and early summer are moist and warm, and soil animals thrive near the surface. To escape the extremes many soil animals make their way to lower levels of the soil and take their food supplies with them from the surface. As they digest their plant food they release the humus that provides the black colour. These fertile chernozem soils are extremely important because they are the soils of some of the world's chief grain-growing regions.

Figure 10.9 A chernozem or black earth of Canada
Notice the deep organic-rich surface layer.

In upland areas   *Upland soils*

In upland areas soils experience more precipitation and lower temperatures than in lowlands. As a result much more water soaks through the soil and carries away vital nutrients. Low temperatures do not suit soil animals very well and, with fewer of them to decompose the dead plant leaves, vegetation

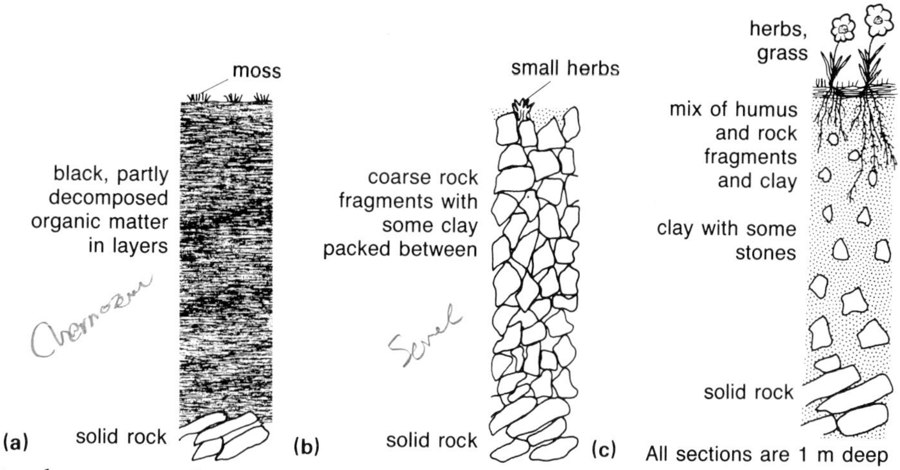

Figure 10.10 Some surface materials

Figure 10.11 Profile of a podzol soil
This is a podzol soil from an upland area. The clays and nutrients have been washed out of the upper layer, leaving only ash-grey sand. The organic material rests *on top of* this and it is *not* mixed with it. Conditions are too acidic for earthworms and most other soil organisms to survive.

(a) moss / black, partly decomposed organic matter in layers / solid rock — *Chernozem*

(b) small herbs / coarse rock fragments with some clay packed between / solid rock — *Sorel*

(c) herbs, grass / mix of humus and rock fragments and clay / clay with some stones / solid rock

All sections are 1 m deep

tends to accumulate on the soil surface. In extreme cases such as in muskeg swamps or very wet valley bottoms the accumulation gets so thick that we use the term **peat** to describe it.

Once the number of soil animals decreases, a vicious circle of events begins and a new type of soil called a **podzol** develops (Fig. 10.11). The word podzol comes from the Russian word for ash, and an ash-coloured layer is very clear.

With needles building up on the surface, the nutrients needed for further plant growth are not returned to the soil very quickly. Those nutrients that are released are often washed through the soil by the high amount of precipitation. Most plants cannot tolerate cool wet conditions and a lack of nutrients. Only woody plants such as junipers and coniferous trees survive because they can store nutrients effectively. However, the resinous nature of these plants makes them difficult for the few remaining soil organisms to decompose, so the dead conifer needles tend to build up on the soil surface even more. Rain water falling on this poorly rotting vegetation becomes much more acidic and this passes through the soil, removing even more nutrients. Few soil animals like acid conditions, so fewer soil animals inhabit the soil. In this way the vicious cycle of soil decline continues.

The extreme acidity of these soils is clearly seen in the colours of the layers (Fig. 10.11). The surface layer is black because it consists of an accumulation of poorly rotting leaves. Rain water passing through this surface layer becomes acidic and it leaches away (washes out) the nutrients

Figure 10.12 A podzol soil

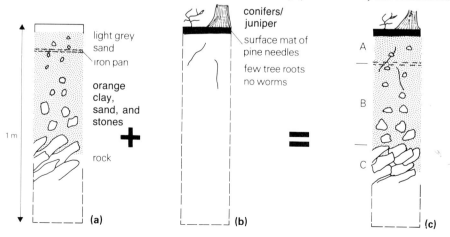

and the iron from the layer below. Without the iron to stain the soil it becomes a distinctive ash-grey colour. Below this layer the iron is redeposited because conditions are less acid. The extra supply of iron turns this layer a rusty orange colour. Podzols are easily recognized by their coloured layers: black, ash-grey, and orange.

The role of relief

Getting it together

Soils are closely influenced by climate, drainage, vegetation, and the time during which they have been forming. On a worldwide scale, climate may be the most important factor influencing the type of soil that develops, but on a local scale this is likely to be less significant.

A valley often contains a sequence of soils, even if it has been eroded in only one type of rock (Fig. 10.13). On the upper part of the valley, rain water may soak through the soil and leach away nutrients. In this way a podzol soil may develop. Further down the valley, nutrients are also leached away, but they may be replaced by those from up slope. Slopes are also more sheltered, temperatures higher, plants grow more quickly, and more soil organisms are present to decompose and recycle plant tissue. Brown earths are more common in these lower areas. Finally, at the bottom of the valley, near the river, the water that has soaked through the soils builds up before entering the river. Here there is very little leaching, but the water may remain near the surface for longer periods. Water-logging turns these riverside soils dark blue-grey in colour.

Figure 10.13 A sequence of soils in a valley

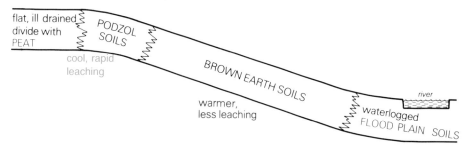

SUMMARY: SOIL FORMATION AND MAJOR SOIL GROUPS

Soils develop on unweathered rock. This may be solid rock, or transported fragments such as river alluvium or deposits from ice sheets. Chemical weathering attacks this parent material (Fig. 10.5a) and forms a weathered surface layer of clay particles. Weathering also releases the nutrients that plants need. Plants and soil creatures now invade the soil. Micro-organisms decompose any dead plant tissue, producing humus and releasing nutrients so that more plants can grow.

Slowly the incorporation of organic matter into the surface layer turns it a darker colour (Fig. 10.5b). Chemical weathering proceeds below this A layer, helped by acids released as plants are decomposed. Soil micro-organisms live only in the upper soil where air and new plant material are available. The deeper weathered layer therefore contains little organic matter. It is called the B layer or subsoil (Fig. 10.5c).

Brown earth soils have a dark brown A layer and a medium-brown B layer, with a gradual transition between them. **Podzols** have been heavily leached. They are acid soils and they contain few soil creatures to mix organic and mineral material. As a result they develop sharply contrasting layers. Only the A layer contains organic matter and rests directly on the surface; the layer below is ash grey in colour due to extensive leaching; the B layer is orange where iron has been redeposited. Some podzols have all the iron redeposited as a natural concrete called **hard pan** at the top of the B layer. **Chernozems** are deep black earth type soils of the plains. Their deep black A layer is partly the result of soil creatures migrating up and down the profile during the year because of extremes of heat and cold. **Tropical red soils** (formerly often called laterites) are deep brown earth type soils where the iron has turned a red colour under tropical conditions.

11 Clothing the surface

Ecosystem principles

The delicate web

The earth has several "skins": below our feet is a skin of crustal rocks into which the landscape has been carved; above our heads is a skin of atmosphere which provides our weather, and the water and wind that help erode the land; but between them there is a delicate and beautiful skin of green — the vegetation that clothes most of the earth's surface. This web of plants is vital to our well being for it provides food for people and animals. It also holds the soil in place and, in so doing, exercises a control on the landscapes far out of proportion to its apparent size and strength.

People have had more influence on the distribution and character of plants and animals than on any other part of the earth. Consequently, we have to make even greater efforts to sort out the natural order of things so that we may best learn how to preserve some of the remaining natural environment while adapting much of the rest for our long-term survival.

The nature of vegetation

From ancient beginnings

The patterns of plant and animal life on the earth's surface are the result of long periods of evolution and adaptation to the natural environment. Some rocks in Canada, for example, show fossil forms of life over 3000 millions years old whose descendants can still be found building the limestone of the Great Barrier Reef of Australia.

Most scientists think that life probably formed at a very early stage in the history of the earth when the atmosphere was very different from that of today. At this early stage there was a lack of oxygen, but the atmosphere was just right for the most primitive forms of life to be created. In a research institute in England, scientists are beginning to succeed in recreating these conditions in a laboratory. Using a mud "soup" of the kind that may have existed on earth near volcanoes, these basic building blocks of life can be formed. This may show the way that life and evolution began. Once started from these simple substances, the long path of evolution has been able to produce the millions of different life forms that we see today.

All living things are made from very simple building blocks organized into very complicated arrangements. The basic chemicals that make up our bodies, for example, could be separated and put into just a few jars on a shelf! One jar would be labelled calcium, another iron, and so on. Below the jars would be a large tub of water. As with all living things, human bodies are mostly water.

But why do we need to understand that our origins may go back to simple forms created in hot mud pools? What has this to do with geography? First, it allows us to see that only simple materials are needed for life. Secondly, we can see that a vast length of time has elapsed since the starting date. In this huge time span there should have been plenty of opportunity for development, specialization, and adaptation to produce a wide range of life forms, each adapted to its own environment. It is the geographer's task to explain the present distribution of plants and animals over the earth's surface. So geographers need to know something of past events before trying to understand the present.

Habitats

Home sweet home

Every plant and animal has a **habitat** — a home — that is suitable for its survival. This is the fundamental idea that underlies biogeography. But plants and animals do not live in isolation: one may provide a shelter for another, some may provide the food for others; and so on. An environment in which a group of animals and plants all live together in a stable and mutually dependent way is called an **ecosystem.** Ecosystems (Fig. 11.1) provide frameworks for studying communities of plants and animals along with the local climate and soils on which they depend.

When we look around we see that most places provide a habitat for some living things. Even the top of a mountain, a desert, or a piece of wasteland in a city has its own distinctive ecosystem. This tells us that life has become very well adapted to a whole range of environments. Now we need to know just how this has become possible.

Natural selection

Little by little

It is probably just as well that we do not all think, feel, and look alike, but we do still have much in common. Together, humans make up a **species,** a biological group that is capable of interbreeding.

Over the past half century, archaeologists have excavated many sites where our ancestors lay buried and they have found considerable differences between these people and ourselves. Several million years ago people had to

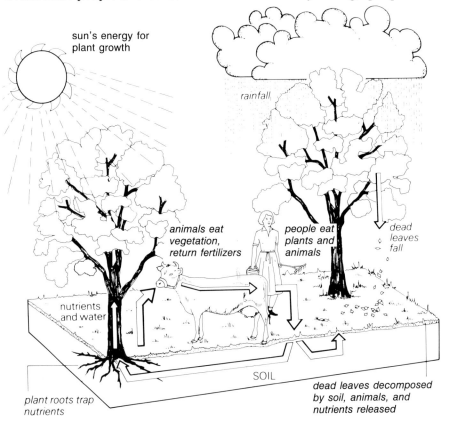

Figure 11.1 Ecosystems A natural ecosystem provides a stable habitat for many species.

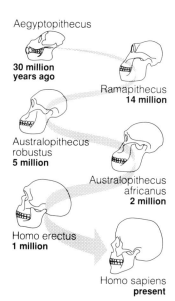

Figure 11.2 Evolution of Homo sapiens Humans have passed through many stages in the past 30 million years.

Food chains

have teeth that would crush food rather than bite, but as this became less important the large teeth in the front of the mouth became smaller and developed sharper edges. Many other changes can also be recognized: the shape of the head, the height of the body, and the amount of hair have all changed. Reconstructed models of prehistoric skulls show that they looked very different from ours and, yet the changes by which they evolved came about only by generation after generation of tiny, repeated differences (Fig. 11.2). If the differences between parents and children were an advantage, then they helped them to survive and to be more successful in their hunt for food, and increased their resistance to disease. If the differences were a disadvantage, then they would be less likely to survive. In this way the more successful trends in evolution have been preserved while the others have died out. This is how **natural selection** works to help species adapt to their environments. It is often called the "survival of the fittest."

Homo sapiens is the most highly evolved of all species and they can survive in the most varied environments. Most living things are not as mobile, nor can they go in search of food and shelter. Plants cannot move with the changing seasons and may have to brave harsh, frosty winters as well as hot dry summers. So, in order to understand the distribution of a species we must find out what it needs to survive.

Eat and be eaten

One of the most basic needs for most living things is sunlight. Plants, for example, use water, oxygen, carbon dioxide, and nutrients as they grow, but they need the energy of the sun to combine these substances into new tissue. Most animals eat plant tissue in order to convert this stored energy into forms they can use in their turn. Thus one creature is dependent on another for survival. Moose, for example, are herbivores (plant eaters) which provide food for wolves which are carnivores (meat eaters). The plants, moose, and wolves are linked in a food chain. At each stage along the chain only a small part of the stored energy is passed on, so wolves need to eat a lot of moose in a lifetime and the moose need to eat a large amount of

Figure 11.3 A food chain

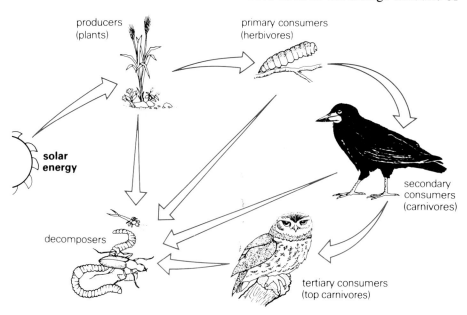

plant tissue (Fig. 11.3). However, if there were too many moose in a region, the plants would not be able to grow fast enough to replace what was eaten, overgrazing would occur, and the plant supply would run out. Eventually the moose would starve and some would die until their numbers were in balance with the rate of plant growth. In a stable ecosystem, therefore, there must be the correct balance of carnivores and herbivores so that links in the food chain do not get out of step.

Each ecosystem must be adjusted so that every part of it is in equilibrium in other ways too. For example, some plants can flourish in the presence of one another because they do not compete for the same amount of light or the same type of nutrient. Direct competitors are seldom found together in equilibrium.

Many types of stable ecosystems are possible, depending upon the climate, the slope of the land, and the soil. On a worldwide scale such ecosystems often give rise to distinctive types of vegetation. Many of these have been given common names such as the Prairie grassland and tropical rain forest. However, within these broad groups there are smaller, but distinctive groups that adapt to, for example, a wet riverside or a sandy soil.

The range of species

Far and wide

We call plant and animal tissue food, but to plants "food" comes as a drink sucked through straws (the roots). An appetizing drink to a plant is one containing all sorts of "goodies" such as calcium, nitrogen, and sulphur. Except for some dissolved metals that are poisonous, the general rule is the more dissolved substances from rock and decayed organic matter in the water, the better. In fact, plants are mostly indiscriminate feeders — they suck up nutrients more or less in the proportions that are available in the soil water, and the main control on plant growth is the nutrient that is in shortest supply. This is the growth "bottleneck." So here we find a potential source of limitation for some species that may require a copious supply of one particular nutrient or loathe the presence of another. Rhododendrons and junipers, for example, just will not grow well in calcium-rich water, which is why they are found in sandy acid soils. Alfalfa and clover, on the other hand, will not grow without a lot of calcium!

Figure 11.4 Colonization and Succession
These ferns and shrubs are growing in a pool of sunlight where some trees have been removed. Given sufficient time, the vegetation will regrow to the original form and composition of the forest.

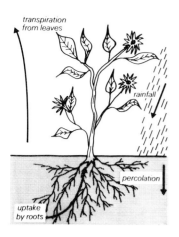

Figure 11.5 Limits to growth
Many plants need continuous water supplies to provide enough nutrients for fast growth. Where water is limited, plants tend to grow more slowly and adapt to conserve their supplies.

Very few of us prefer winter to summer, and most of us feel better on a sunny day than on a dull one. So it is with plants. Indeed, plants need a certain amount of sunlight to grow at all. In many cases where there is too little light, either because of distance from the Equator or a canopy of tall trees, the seeds of some species simply do not ripen or germinate. As a result their reproductive cycle is not completed (Fig. 11.4). Remember that, even with 24 hours of sunlight in the arctic summer, this low-angled radiation cannot provide the same energy as just 12 hours of tropical sunlight. This does not mean that shady or polar areas are without vegetation, but rather that different plants need different amounts of energy, and only plants tolerant of low-energy levels can thrive in these difficult places — an example of "adapt or die."

Both sunlight and nutrient supplies are important in controlling the distribution of species. But looking back to our list of plant needs we see that the other vital requirement is for water. So what happens if the water supply is reduced? In fact, water does much more than just provide a means by which nutrients can reach plants: it takes part in the formation of tissue and, just as air fills out the shape of a balloon, it maintains the shape of non-woody parts of tissue. A glance at a dried plant will soon confirm this last point.

Unfortunately, plants cannot just blow themselves up with water and then remain inactive. They need to keep pumping water so that the "food" of nutrients and dissolved gases can be fed in. This constant need for more water means that the "used" water has to be removed. This usually happens as a chain reaction started by sunshine or air drying the surface of leaves. More water flows to the surface of the leaves and this in turn pulls water up from the roots (Fig. 11.5). And if there is a limited supply of water, some rather special adaptation is needed, otherwise the continued removal of water from the leaves will simply cause the plant to wilt and die. Small woody plants or those with thick rubbery leaves are typical adaptations (e.g. the olive, cactus (Fig. 11.6), acacia tree, and mesquite bush (Figs. 11.6, 11.7).

Figure 11.7 Plants of the desert

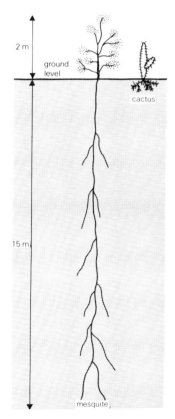

Figure 11.6 Adaptations
The mesquite bush survives during a drought by tapping reserves of water up to 15 m below the surface. The cactus, without deep roots, has had to adapt in a different way.

Topic Plant resistance to fire, drought, and pests

In the deserts of North America, one of the most commonly found bushes is a grey-green woody plant with small leaves, called mesquite (Fig. 11.6). In these deserts, precipitation occurs only a few times a year and there are long periods of drought. At the same time temperatures can reach well above 30°C — enough to make most plants wilt in minutes! Yet the mesquite grows on and survives because it has adapted to its environment in a remarkable way. Although woody stems and small leathery leaves are important in reducing water loss, the key to the mesquite's survival is seen if one is dug up. To dig up a mesquite complete with roots may require a mechanical digger and pneumatic drill! It is normal to find the tap root sinking down to 15 m — more than the height of a house — just to find a water supply.

When the famous wooden horse was being wheeled through the gates of Troy over 3000 years ago, a small sapling had just begun to rise above the forest floor in California. The seed from which it grew weighed less than 4/1000th of a gram. When Christ was born, the tree had grown to over 30 m in height. Today it is 90 m tall and weighs 1320 t (Fig. 11.8). Such is the history of the giant redwood and, with a 3000-year growth period to its credit, it is surely one of the best survivors. However, its means of survival are quite different from the mesquite. Because it takes so long to grow it must be very resistant to disease and for this purpose it has grown a bark up to 60 cm thick. The same bark has enabled it to resist fires that have swept through the forests over the centuries. Indeed the redwood has benefited from forest fires because other species have been burned away, reducing competition for growing space.

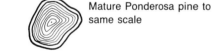

Figure 11.8 The General Sherman Tree
This redwood is thought to be the largest living thing on earth. It has lived for over 3000 years and weighs about 1320 t.

Sunlight, water, and other nutrients are all fundamental to the growth of plants. Of these, water and sunlight are the most critical and, because these are also features of climate, people have long associated vegetation zones with climatic zones. In a broad sense there is obviously logic in this but, as we have seen, on a more local scale there can be many other controls on the plants found at any one place.

Let us leave this idea of broad controls by reemphasizing the effects of water restriction. When water is limited, the number and size of plants that can be supported in a given area decrease both in volume and range of species. Eventually the plants may have to be widely spaced to obtain enough water for themselves. Water has also had such a powerful influence on evolution that widely separated areas containing unrelated species can take on much the same appearance and pattern.

Exercise 11.1 Desert survival

Figure 11.7 shows an area where precipitation is very low and irregular.
1. Explain why plants have to make special adaptations to hot, dry conditions.
2. Choose two different types of desert plant and describe how each plant type survives.
3. From the photograph, list any other adaptations that you can see.
4. Why do the plants not cover the ground completely?

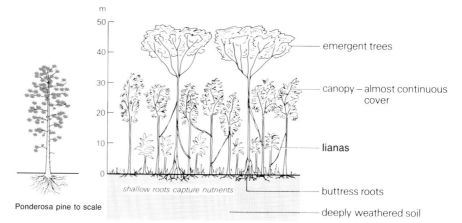

Figure 11.9 The vegetation pattern of the tropical rain forest

Figure 11.10 Tropical rain forest buttress tree

Climax vegetation

Survival of the fittest

Let us start an examination of vegetation by investigating the rain forests of the Amazon, Zaïre, or Southeast Asia where there is much warmth and precipitation. In these places we can try to see what the results of a long-term battle between species are like.

A **tropical rain forest** is dominated by giant trees, some over 45 m high (Fig. 11.9) whose roots are even adapted into "buttresses" (Fig. 11.10). These trees emerge above the main tree species which provide an almost complete canopy of leaves at about 30 m and block out much of the sunlight from the layers below. Hanging from the trees are the lianas — creeping plants, which like the ivy, live on other trees. These are the natural ropes of the jungle that some animals use for swinging from branch to branch.

Below the tree leaf canopy there are smaller trees that can grow with less light, and then below them, on the forest floor, are the shade-tolerant plants that need low light, high humidity, and constant temperatures to thrive. Here is a mature and complete stage — a **climax vegetation.** In a tropical rain forest there are more species of plant and animal than anywhere else in the world. There are evergreen trees and some deciduous trees, shrubs, herbs, and mosses all surviving together.

However, the equilibrium is a continual battle for survival. You can see how fierce the competition is in such a forest when a giant tree dies. Often it will fall over and tear out other trees as well so that a broad swathe of forest is exposed to the full light of day. This is the time when the vegetation changes rapidly and the battle for colonization of this new empty space begins in earnest. First, grasses and other low, sun-loving plants colonize the ground. This light-tolerant ecosystem is not stable, however, and it is soon overwhelmed by a scrubby bush-dominated ecosystem. But even this dense thicket-type of vegetation survives only until the saplings of larger trees grow to a sufficient height to shade the bushes. As the shade increases, the light-loving plants are deprived of their source of energy and they die back, and their places are taken by shade-tolerant plants. In this environment of ample heat, sunlight, and moisture, the sequence of changing ecosystems is very rapid and it may take only a few years before large trees dominate.

The dense vegetation that builds up in these early stages has led to the term "jungle" being used for parts of the rain forest. In such areas it may not be possible to see ahead for more than a few metres, and progress on foot requires constant cutting of trails. Indeed, people often describe the nature of the rain forest by the distance that can be seen ahead, for example, a 10 m jungle or even a 2 m jungle! However, gradually the largest trees grow through the newly colonized jungle and then provide shade, so that many low tree species die back and the forest becomes more open again near the ground.

So the key to all climax vegetation patterns is the survival of the most suitable *in the long-term*. Where heat, light, and moisture are abundant, as in the tropical rain forest, it is possible to support a great bulk of plant matter at every level and to maintain a wide variety of species. However, away from these regions of plenty, water limitation and/or temperature limitation are encountered and the plant bulk is reduced. Under these conditions different species come into their own and dominate the ecosystem.

Topic Plant colonization in a garden

Suppose a piece of land is cleared by a landslide or a volcanic eruption, or as a building site. Initially the ground is laid bare, but it doesn't remain that way for long. If the piece of bare ground were a cleared building lot in Canada, the first colonizing plants would probably be called weeds. Our gardens are also cleared ground and a battle is engaged through the summer between weed and gardener. This is the main reason gardening takes up so much time.

Suppose a garden were abandoned and left to these weeds. Dandelion is a typical example of a Canadian weed, which will rapidly grow over abandoned gardens. It is soon accompanied by many other colonizers such as thistles, knapweed, and grasses, until the whole surface of the ground is covered once again. After a while the odd shrub, often some form of maple or alder may invade, although eventually trees such as birch and aspen will grow above this ground cover.

Trees create shade. Some of the initial ground cover may not be able to live without direct sunlight and so it will die. In this way competition ensures that the composition of the vegetation changes. It may take up to 100 years for a proper forest to develop over the garden, but even then a stable plant community will not have been achieved. The 'pioneer' trees like birch have only a short life and after a couple of hundred years they will be replaced by longer-living species such as douglas fir. Only then will a more or less stable or climax vegetation be attained over what used to be a garden. You have only to look at disused railway lines, factories, or airstrips to see colonization in action.

People's use of plants **Plants create patterns**

A climax vegetation is broadly dependent on climate in most cases. For example, climate usually determines whether the vegetation will be of tundra shrubs and mosses, coniferous forest, savanna parklands, and so on. Nevertheless, within this broad framework the detail is largely under the guiding influences of soil, drainage, and slope. There are, of course, some instances where climate may not be the main control. For example, a closed depression from which the water cannot escape may well develop into a swamp or bog, and permit only water-tolerant plants such as sphagnum moss and rushes where otherwise there might be forest.

Bearing the local differences in mind, in broad terms similar climates around the world will encourage similar types of climax vegetation (Fig. 11.11). Not that the same species will be present in the Zaïre jungles of Africa as in the Amazon jungles of South America, but the overall tropical rain forest will look the same. This is important for people because it means that they can often take a plant from one region of the world and grow it successfully in a different region with a similar climate. There are numerous examples of the benefits this has brought: grape vines from America have been grown in France to improve the disease resistance of European stock. Coffee trees are now grown in Asia, West and East Africa, and Central and South America although they originally came from Ethiopia in northern Africa.

Nowadays people tend to ignore or fight the resilience of the biological world in its attempts to restore itself. Farmers resort to extensive use of herbicides every year in order to prevent "weeds" colonizing their fields. Yet some people still use nature to advantage. The foresters of Scandinavia and Canada encourage reforestation in their lumbering areas as a means of renewing their reserves. Loggers cut trees for baseball bats, hockey sticks, and fence posts knowing that in the struggle toward a climax vegetation the trees will grow up again. Slash-and-burn farmers in tropical forests know that they can cultivate land for a while, then leave the forest to grow again, gradually replacing the needed soil nutrients as it does so (pages 140 & 151).

Figure 11.11 Generalized major natural vegetation zones of the world

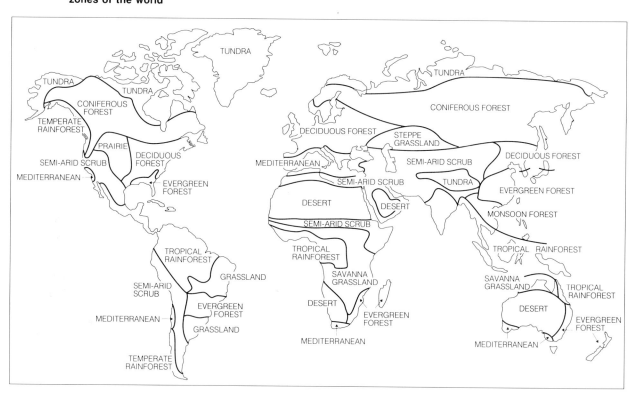

Topic Major natural vegetation forms of the world

Tropical rain forest. Tropical rain forests occupy about 17% of all the land surface covered by broad-leaved trees. As described on p. 127, rain forests may contain hundreds of different tree species, sometimes occurring in patches, sometimes singly. Because there is no cold season there are trees in all stages of their reproductive cycles at all times of the year. Intermixed with a bewildering variety of evergreens (Fig. 11.12), there will be some deciduous trees coming into leaf at the same time as leaves are falling from others, and yet others are in flower.

Each type of tree grows to a different height, so that a forest has many tree canopy layers (Fig. 11.9). This layering has given the rain forest the name "gallery forest." Together these tree layers shade the ground so that near the forest floor there is little light. Few plants will grow in deep shade, so it is often possible to walk about quite easily.

The tropical rain forest is a closed system — an ecosystem with a very tight control over its nutrients. Luxuriant forests can be found even on soil that is not very fertile because the trees recycle nutrients so effectively. The trees are shallow rooted and they trap nearly all the nutrients as they are released by rock weathering and the decay of dead leaves. These are used *immediately* to make new growth. It is a never-ending cycle but *it relies on the ecosystem remaining undisturbed.* So delicate is the balance within the cycle that, once disturbed by clearing of trees, rain soon washes nutrients from the soil; in the worst cases, forests cannot recolonize severely degraded and eroded tropical soils. Rain forests occur in the Amazon Basin, Mexico, Zaïre, and Southeast Asia, all regions which are being exploited at a rapidly increasing rate. The possible effects of this are causing great concern (see p. 153).

Savanna. Savannas are tropical grasslands with a small proportion of trees. Together, grass and trees make up a sort of parkland, covering some 20% of the world's land surface. Most savannas occur between tropical rain forests and semi-arid scrub that border deserts. Because they are so widely distributed, the type of grass varies considerably from the African elephant grass, which is several metres high, down to low sparse grasses on the borders of the deserts. Trees in the savanna all have deep strong roots because the precipitation in these regions is seasonal and there are long dry periods. Trees become specially adapted to these conditions in other ways as well. Many, such as the acacia (Fig. 11.13) are deciduous, shedding their leaves in the dry season to conserve water. They also often have a thick bark and scaly buds to protect themselves from the fires that become a major hazard in the dry seasons.

It is difficult to know how natural the savanna parkland really is, because people have burned and cultivated these areas for so long. However, in some regions woodlands are more common, so it is possible that the natural vegetation of the wetter savanna should really be deciduous woodland (Fig. 11.14).

Figure 11.12 Rain forests
The canopy of mature tropical rain forest can be so dense that the ground is totally obscured.

Figure 11.13 Savanna parkland, Sudan
Above. In the dry season and, *below,* a little after the start of the wet season.

Steppe and Prairie grasslands. Steppe and Prairie grasslands occur in temperate regions mostly on the poleward side of the great deserts. Like the savannas they may owe their vegetation to the low precipitation and a long dry season which restricts most tree growth. Steppe grasslands have an additional problem to overcome because they experience very cold, snowy winters.

Figure 11.14 Savanna
This is a semi-natural area in Zimbabwe, where low-growing acacia trees make a woodland with grass patches. The vegetation and soils are developed on a granite batholith.

Figure 11.15 Deciduous woodlands

Summer

Winter

However, the grassland that we see today may be as much an effect of an enormous population of grazing animals. For example, 50 million bison used to roam the Prairies, each eating as much as a cow does. Again we do not know how "natural" the steppe, pampas, or Prairie grassland really is. Much of it anyway has long since disappeared under the plough to become some of the world's great wheatfields (see p. 143).

Temperate deciduous forests. As precipitation becomes more regular and plentiful, the steppe and Prairie grasslands sometimes merge into deciduous forests. Deciduous forests of the mid-latitudes (Fig. 11.15) consist of only a few dominant trees, such as oak, maple, aspen, and hickory, in contrast to the amazing variety of the tropical rain forests. Light often filters through the tree canopy and allows an undergrowth of smaller species such as alder and a ground vegetation including bracken, thistles, and grasses. Temperate areas receive less energy from the sun than do areas nearer the Equator, so growth is slower and trees less densely packed. Nurtients are also recycled more slowly and they may accumulate on the soil surface for many months before being decomposed.

The North American and Asian deciduous forests have many more species in them than do the European forests. This is because trees in Europe had to migrate from Turkey around the difficult barrier of the Alps since the ice retreated only 12000 years ago. There is no such barrier in North America or Asia (Fig. 11.21).

Coniferous forest. Conifers have adapted to a wide range of harsh environments, including dry and wet sites. The trees are all very efficient in their use and storage of nutrients, because in these areas nutrients are in short supply. Spruce is prominent in areas where precipitation is heavy or where soils are wet. Pine, with a shallow

Figure 11.16 Sand, salt, and xerophytes
Few hardy xerophytes (drought-loving plants) survive in an area of central Australia which has low, unreliable precipitation and high evaporation. Flash floods from infrequent storms soon give way to salt pans in basins of internal drainage.

rooting system, tends to be found more often in drier sites. Together with other species the coniferous forest stretches in a broad belt polewards of the deciduous forests in Europe. However, in the drier areas of the Americas and Asia there are no deciduous areas, and the steppe grassland merges directly with the conifers. Conifers also occur on mountain sites as far apart as the Rocky mountains and the Alps of Europe.

These forests are made of even fewer species than the deciduous forests and they often appear dark and monotonous (Fig. 11.17). Here, recycling of dead plant matter is very slow. The floor of a coniferous forest is always littered with a thick layer of needles that have not yet been decomposed. The soils are almost always poor in nutrients and the number of soil creatures is low. These forests have rarely been cleared for agriculture because the cold climate and poor, pebbly soils make them unsuitable for most other plants. They do, however, provide the world's largest timber reserves and are widely exploited in Canada, Scandinavia, and the Soviet Union.

Tundra. The poleward or altitudinal limit of the coniferous forest is rarely sharp. As the boundary is reached, trees become more stunted and eventually reach only the height of shrubs. The problem that all plants face in these areas is survival under extreme climatic conditions. Arctic plants face long winters and poor summer heating and so are very restricted in size and variety. Many seek the shelter of rocks and hollows for their survival. Most alpine mountain tundra is more varied because it has a shorter winter and greater summer heating. Therefore, don't expect the variety of pretty flowers found in the Alps to carpet the regions of the Canadian Arctic or Siberia. The most common arctic plants are cotton grass, sedges, mosses, and lichens which together provide a grey-green cloak of just a few centimetres in height to these exposed regions of usually water-logged soils (Fig. 11.18).

Figure 11.18 Tundra vegetation
Tundra plants are low shrubs, mosses, and lichens, and they give a grey-green look to these otherwise barren lands.

Figure 11.17 Coniferous forest
They are often monotonously regular.

Figure 11.19 Desert with xerophytes
The red sand dunes of central Australia, near Ayer's Rock, have been stabilized by xerophytes (drought-loving plants) like spinifex grass and mulga trees.

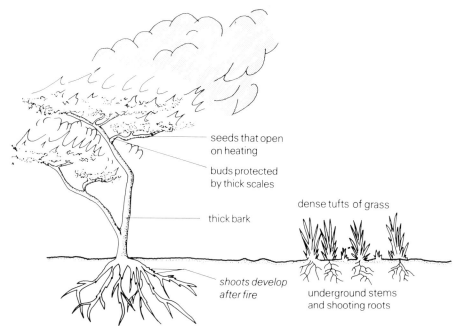

Figure 11.20 Fire-resistant plant features

Exercise 11.2 Drought and fire

1. Figure 11.19 shows xerophytic plants growing in an arid region. Suggest how they may have adapted to survival in low precipitation regions.
2. Figure 11.20 shows some of the features of vegetation that might help the plants survive fire. Explain how each of these is useful. How are fires likely to be important in determining the nature of (a) evergreen rain forest, (b) steppe, (c) savanna, (d) temperate woodland, and (e) desert?

The effects of continental drift

Drifting apart

In earlier sections we have seen how plants have a particular set of environmental needs and how they fit into a community of other plants to make up an ecosystem. But we have also seen that continents have drifted in the past and are still drifting today. Only 50 million years ago, British Columbia was experiencing a tropical climate. It has been only 12 000 years since an Ice Age left the land barren, and yet it is now clothed in vegetation.

These two examples show that there have been some very remarkable changes in vegetation in the past.

However, plants cannot simply pull up their roots and move to more suitable areas. Change happens because conditions become too harsh and species die off in the direction of climatic worsening, and at the other end of the range the environment becomes better and some seeds will grow for the first time. This change of distribution is called **migration.** It must have happened in the past and, of course, it is still happening today.

There is one other important factor to be remembered when thinking of migration. As continents drifted across the earth they did not necessarily stay in one piece. Continents that have split in the past have been out of contact and their **flora** (plants) and **fauna** (animals) have thus evolved separately. Given a long enough period of time new species will evolve. This is why much African and American flora and fauna are distinctively different in detail despite having the same basic shapes and functions. Africa and South America both have rain forest and grassland belts and they look alike from a plane, yet from ground level it is clear that there are considerable differences in the species present.

Migration patterns

Silent waves

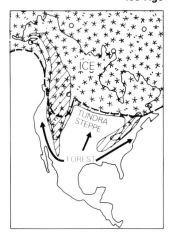

Figure 11.21 The spread of vegetation at the end of the Ice Age

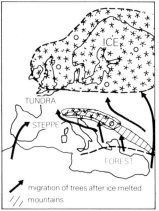

We can show that the pattern of plants and animals is always changing if we answer the question "What happens as the climate improves over a previously inhospitable area?" To answer this question, we can examine western Canada's vegetation over the past 10 000 years.

As the ice retreated from the land it revealed a barren wilderness of crushed and shattered rock. This was the foundation of our soils, and the home of colonizing plants and animals as they made the first tentative steps from the areas south of the glacial limits. Watching the colonization of Canada must have been much the same as observing the changing pattern of vegetation from the North Pole to Oregon would be today. However, instead of making a journey across a quarter of the globe, Native tribes may have noted the different forms of vegetation move past them (Fig. 11.21).

Today, the vegetation that first grew on the barren landscape can be found only on high mountains or within the Arctic Circle. In these areas an ecosystem including mosses, lichens, small flowering plants, and dwarf shrubs make up the pattern called tundra. This ecosystem survives as a major vegetation form in these areas because the plants can endure the rigours of arctic winters with their long periods of snow cover as well as the water-logged soils of summer. No other vegetation forms can withstand this type of climate, but in warmer regions what chance would such low-lying vegetation have against the trees of a forest? Our first wave of vegetation may have been of tundra plants, but they were soon overwhelmed by the advancing armies of birch, pine, and willow trees in all but a few special places, such as the highest mountains of the Cordilleran region.

For example at Cathedral Lakes Provincial Park in British Columbia, the northeastern edge of the Cascade Range contains spectacular peaks, cirques, and valleys. It is a summer paradise for hikers who appreciate its magnificent scenery and the beauty of its sparse but complex and colourful

Figure 11.22
Cirque Lake, Cathedral Lakes Provincial Park, south-central British Columbia

plant life. Much of the area is so steep that little of the soil, developed over the past 10 000 years, remains in the cracks of the frost-shattered rocks and between the angular boulders on the scree slopes. Delicate mosses, lichens, and flowering plants live precariously in the arctic environment. The plants flourish in the brief summer, but their rate of growth from year to year is almost imperceptible. Their recovery from damage is still slower. For that reason, it is not easy to gain access to the park. A 19 km hike, which involves a 2000 m climb, takes you to the park lodge. There, you are asked to "take nothing but photographs; leave nothing but gentle footprints."

Coniferous trees – pines, firs, spruces, larches, cedars, and hemlocks – now dominate the vegetation of most of Canada. As warmer conditions spread over North America, Europe, and Asia after the end of the last cold period, seeds from the hardy birch, willow, and coniferous trees, such as spruce and pine, were able to grow among the tundra mosses and lichens. Soon they developed into mature trees and shaded the ground so that the tundra vegetation died. As these trees matured, they scattered their seeds farther north and continued to migrate.

In amidst the conifers, seeds of deciduous trees started to grow, but they matured more slowly and, therefore, spread less quickly. However, eventually the new arrivals of aspen and, later, willow and maple became more successful in their turn and replaced the conifers in the most climatically favourable areas.

Natural plant succession is a very slow and uneven process. However, what occurred was a series of waves of vegetation advancing across Canada as conditions became warmer. First came tundra plants, then the conifers and, in the most favourable areas, the deciduous trees which now cover most of southwestern British Columbia, the St. Lawrence lowlands, and Atlantic Canada. The dryness of the Prairies discouraged tree growth; instead a complex grassland ecosystem developed over most of Alberta, the southern half of Saskatchewan, and the southwestern corner of Manitoba. The highest levels of the mountains and the most northerly regions of Canada still support the original tundra plants. In every area, except the tundra, each change in the plant community has made the ecosystem more complex since every new group of plants dominates, but does not entirely eliminate the original vegetation.

Exercise 11.3 Mountain climates and mountain vegetations

1 Figure 11.23 shows a profile from a lowland plain in a warm dry region, such as California, to a mountainous peak. Using your knowledge of the effects of mountains on precipitation (from ch. 5, p. 60) describe how the precipitation would vary between the plain and the mountain peak.

2 Explain how precipitation changes might account for the changes in vegetation.
3 Describe the vegetation you would expect to find between the tree line and the snow line, and explain the special characteristics of the environment that allow it to survive.

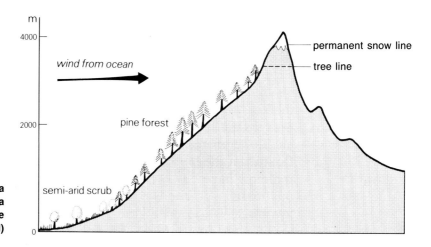

Figure 11.23 A profile from a lowland plain to a mountainous peak (latitude approx. 35°N)

Topic Changes in vegetation with altitude

Mountains show very clearly the way vegetation responds to changes in environment. A walk from an alpine valley up to the summit of a nearby peak provides, for example, a change of altitude of more than 2000 m (Fig. 11.24). The change of climate observed in this climb is the same as a change of over 25° of latitude — the distance from the Alps to the arctic shores of the USSR.

In the valley there would have been a natural forest of deciduous trees, such as beech and oak, intermixed with some conifers. Much of this has been replaced by agriculture, so cereals and permanent pasture are now more common. But as slopes steepen and there is less farming, so the natural forest becomes more dominant. Within 200 m of the valley floor, coniferous forest has completely replaced the deciduous woodland. The main reason for the change is the decreasing temperature and a longer period of snow: this allows a shorter growing season. As the growing season shortens, deciduous trees cannot complete their reproductive cycle — fruits do not ripen and, therefore, trees do not perpetuate themselves.

As we walk up through the conifers it is much darker than the deciduous woodland. Trees are closely packed and little light penetrates. On the ground there is a dense carpet of fallen conifer needles but no ground plants. Climbing higher we find a shorter and shorter growing season and lower summer temperatures until even the conifers become stunted, thin out, and then disappear altogether. We have reached the tree line, which in our alpine valley is 1200 m above the valley floor.

Above the tree line we are in open country for the first time, walking across a carpet of grasses and small alpine flowers that have adapted to survive on cold windswept slopes with a short growing season. However, as we climb to 2000 m only the hardiest forms of vegetation are left: mosses and lichen clinging to rocks in screes.

At 2600 m we reach the snowfield and even in summer all vegetation has now ceased and we are left looking across frost-riven peaks down to the forested valley below.

Figure 11.24 Vegetation changes in the Alps
See also Figure 17.1.

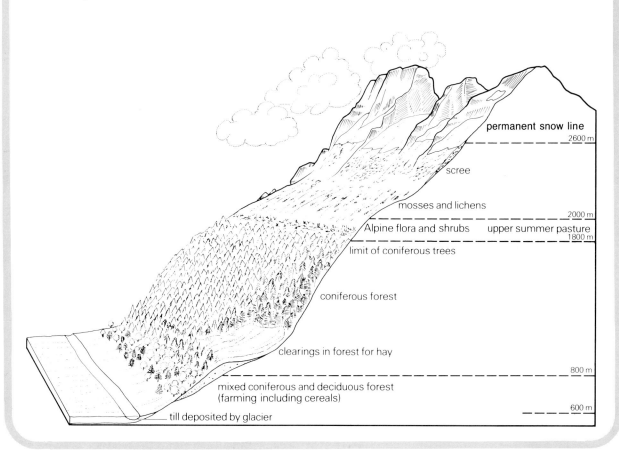

12 Cultural ecosystems

Human impact on ecosystems

In the beginning...

When you look out of the window, how much of the vegetation that you see is really natural and not influenced by people? Can you see any vegetation at all?

Natural vegetation is a tangled mass of plants that you might think had almost been designed to prevent the development of human society. For example, forests make the pursuit of agriculture and the construction of buildings and roads difficult. Similarly, if people want to build a house or a factory they do not usually want a tree growing through the middle of it! Further, most wild plants do not produce much that people want to eat, and even if they do, their productivity is low. Although people have to put up with an inconveniently placed mountain or unpleasant weather, today very few people suffer from competition with natural vegetation.

Weathering, soil formation, and plant growth are all closely connected. For example, rock weathering is speeded up by acids from rotting vegetation; plants need weathered rocks to provide new supplies of nutrients; and soil type is influenced by the plants from above and the rock below. But soils and plants are influenced by people more than any other parts of the physical environment.

People and animals must eat plants to live. Without cultivation and animal farming, there would only be enough food to support a tiny fraction of the earth's population (Fig. 12.1). So, unwanted plants are uprooted or burned so that large areas can be cultivated. In this way people do not simply destroy — they replace. But, of course, the new plants and cultivation methods will also cause the soils to change because the plants with which they were in balance no longer exist.

Clearly, without a careful understanding of the nature of plant-soil relationships, people may eventually destroy the very basis on which their survival depends. This chapter shows some of the ways people have changed the natural order and created their own new ecosystems. We call these ecosystems farmland, pastureland, woodland, and urban land, but because they are cultural adaptations of natural systems they are very unstable and fragile.

Figure 12.1 Low-tech survival In many parts of Africa, people use older methods of irrigation, cultivation, and animal farming. These women in Botswana are carrying water back to their village.

Cultural changes through time

Fire!

One of the world's greatest problems is to feed its ever increasing population. Lack of food is closely associated with the spread of disease and poor resistance to infections, and even today much of the developing world must still face the stark challenge of survival. Although people in the developed world no longer have these immediate problems, in the past they too were often at starvation level because they just couldn't produce enough food with the farming techniques available. For example, in 1348 a plague, called the Black Death, swept across Britain. At the time of its arrival the population was about 4.5 million. Two years later the population had fallen to about 3 million – a third of the people had died. The disastrous effect of the Black Death was mainly because many people were on the verge of starvation. Strange as it may seem, 4.5 million people were just too many to be supported on the land with only simple cultivation techniques and low-yielding crops. Being on the verge of starvation meant that many of the people were prone to disease, just as they are in Kampuchea, Uganda, Mozambique, and other places in the developing world today.

Starving people do not sit around when there is other land to be cleared and cultivated. We can, therefore, be sure that virtually all possible land had been cleared and put under the plough. But as a consequence, the natural vegetation in Britain disappeared very rapidly.

As we have seen this struggle for survival reached a crucial stage in the 14th century, but the natural vegetation had already suffered a long line of blows. Large areas of forest had been cleared as soon as people learned the use of fire. Indeed, we still get frequent reminders of the way fire spreads rapidly. For example, in August 1945, someone carelessly flicked a smouldering cigarette from a motorcycle onto the edge of Highway 3 near Allison Pass in southern British Columbia. The new highway ran along the bottom of a deep, narrow valley, through a dense coniferous forest. It had been a hot, dry summer and the grasses, leaves, and needles by the road were like tinder. Within minutes, they smouldered and caught fire. Fanned by a rising westerly breeze, the fire grew quickly and whisked embers into the dry needles of the conifers. Unfortunately, the early events went unnoticed, and fire fighters were too far away to prevent the fire's spread. The fire became so intense that trees on the opposite side of the road ignited! As hot smoke billowed upwards, a new, violent westerly wind surged through the valley. A firestorm roared through the conifers' crowns on each side of the valley. Although fire fighters battled the blaze for a few weeks, over 30^2 km of mature forest were destroyed — all because of one person's carelessness. The damage is still apparent today. So it does not take much imagination to understand that the great forests of Europe were destroyed by only a few generations of human activity. The clearing of trees to increase the area of farmland and the misuse of fire obliterated most forests early in the Industrial Revolution. The forests stood little chance of regeneration because the seeds and saplings were eaten by domestic animals.

In the ways described above, natural vegetation was under attack by people from a very early stage. Although the easiest areas to clear and work were the thinly forested uplands, succeeding generations cleared more and more land so that by the time the Normans arrived in 1066 the natural forest was already severely depleted even in the lowlands. Only in the Royal Parks did it survive intact. In 1700, after continued felling, it was reported that hardly a tree was left standing within 40 km of Southampton. If this is typical of other areas the removal of natural vegetation must have been more or less complete by the 18th century. Instead of forest, the land was laid out mostly in large arable fields.

Slowly, farming methods changed, but this time the British accidentally allowed the trees to revive. The newer methods required land to be divided into rectangular fields, much after the fashion that we see today. To keep animals within the fields, quick growing trees such as hawthorn were planted and trained into hedges. Inevitably seeds of other trees and plants settled

(a)

(b)

Figure 12.2 The bush fallow system
(a) Rain forest is cleared by fire.
(b) A village is established and a variety of crops grown.
(c) Burned tree ashes are piled up to provide fertilizer.
(d) The area is abandoned and allowed to revegetate. In the foreground a variety of cultivated plants can be seen among recent scrub. Virgin rain forest is in the background.

(d)

in these newly made hedges. Slowly the native trees — oak and elm — rose above the hedges to provide variety in the landscape once again.

During the 18th and 19th centuries in Canada, settlers from Europe cleared the natural vegetation with unprecedented efficiency with the tools created during the Industrial Revolution. Across the lowlands of the St. Lawrence, the hills of the Maritimes, and the delta of the Fraser River in British Columbia, the natural coniferous and deciduous forests gave away to farmland which was separated by small stands of trees. But within the forest remnants, the trees had been selectively logged for the best timber. The grasslands of the Prairies changed less dramatically because the natural grasses were replaced by grass crops such as wheat, oats, and barley.

Increasing soil productivity

Out of the ashes

The history of our changing vegetation is not unusual. People have adapted virtually all of the areas where plants grow naturally, and even some, such as deserts, where they do not. But all new types of land use have one basic need in common: they must have a **fertile** soil. If the soil is not fertile the cost of adding artificial fertilizers may make farming unprofitable.

A soil is naturally fertile if it can provide as many nutrients as are needed for the healthy growth of plants. As we have seen, a natural vegetation slowly gathers nutrients released by rock weathering and then traps most of them in a continuous growth and decay cycle. In this way the demands on the soil may be quite small. Tropical rain forests and northern coniferous forests, for example, grow on soils with very few nutrients.

The difference between farming and nature is that whereas nature can cope with a small reserve in the soil by continuous recycling, farmers cannot. Farmers need to remove the natural vegetation in order to grow their crops. They cannot easily use the nutrients stored up in the tree trunks, so they burn them. After burning, only a small proportion of nutrients are left in the ashes, the rest literally going up in smoke. As a result, the farmers begin with a loss of nutrients and by harvesting their crops, they take away even more: *the nutrient cycle has been broken.* Consequently the soil becomes less fertile, slowly in temperate areas and very rapidly in the tropics.

Loss of fertility in a soil has always been a major problem for farmers. Traditionally, they have had to leave part of the land unused (in fallow) for a few years until it has recovered naturally (Fig. 12.2). Then they plough up or burn off the vegetation that has recolonized the fallow site and they use the land until it is again exhausted. This type of inefficient system is still widely found in the tropical areas of the world today and it is called the **bush fallow system.**

Fertilizers and rotations

Fertile minds

In the 18th century, European farmers started to look for ways of ridding themselves of the wasteful fallow land. European populations were rising, therefore production had to be increased. Gradually and by experimenting they discovered crops need different nutrients and that the soil could be used all the time provided a **rotation** of crops was used. For example, wheat and clover can be rotated without wasting land or reducing fertility, because clover adds extra nitrogen to the soil to replace that used by wheat.

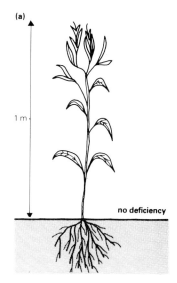

Further experiments showed that if baked limestones (usually called "lime") were scattered on the fields, an increase in yield could be achieved. The age of the artificial fertilizer had arrived. A **fertilizer** is any material that is spread over the soil to restore or improve the nutrient content. Today, farms in developed countries make use of a whole range of fertilizers so that their yields are tens of times greater than those of their ancestors, or even of their contemporaries in the developing world. Each fertilizer has its own effect and the absence of just one element can have a serious effect on plant growth (Fig. 12.3).

Falling apart *Causes of soil erosion*

Artificial fertilizers replace only one of the losses from a soil. They cannot replace the sticky gums that come from humus and which hold the soil particles together. Humus can come only from rotted plants or animal dung.

When soil particles start to fall apart the results can be disastrous. Initially the small particles fall into the soil pores and block them up. Without pores, water cannot drain through the soil properly and it is forced to run over the surface carrying soil particles and starting erosion. When wind blows, it also whisks soil away from bare field with little humus.

The unnaturally fast removal of soil by water and wind is called **accelerated soil erosion.** In the past, Europe did not suffer from this problem because frequent and plentiful precipitation kept the soil moist and prevented **deflation.** Usually light precipitation can even soak into bare soils without running over the surface. However, in other areas to which Europeans took their farming methods this was not the case. North America and Australia provide some clear lessons on the need to observe and try to imitate the way nature cares for its soil. The great soil erosion disasters in the USA, Canada, and Australia tell stories of how people tried to work against nature — and failed.

The Dust Bowl *Erosion in the USA*

The Great Plains are a flat expanse of land which occupies a north-south belt of country just east of the Rocky Mountains in North America (Fig. 12.4). Because they lie in the rainshadow of the Rockies, precipitation is less than 500 mm a year. Indeed, water on the Plains is worth more than

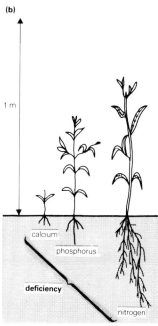

Figure 12.3 Fertilizers Even tiny deficiencies can dramatically affect plant growth and yield. Most fertilizers contain a range of plant foods designed to overcome the loss of growth shown here.

Figure 12.4 Areas most subject to soil erosion in North America

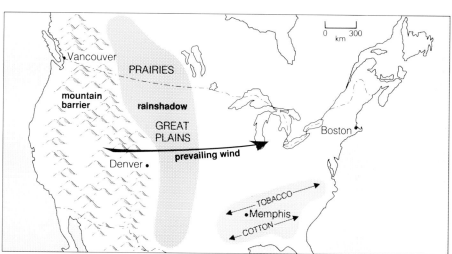

property. In summer the precipitation occurs only as occasional isolated storms, so that in any particular part of the Plains there may be enough rain for good grass and wheat to grow for two or three years; but then there may be a year when virtually no precipitation occurs, when the streams from the mountains dry up and fill with sand. In fact the area used to be called the "Great American Desert" on early maps. In the years with little precipitation, strong winds dry up anything that is exposed on the surface, crops fail, and despair comes to farmers.

Before the European settlers arrived, this was the grazing land for vast herds of bison whose numbers were controlled naturally by the amount of grass and water available. When the railway reached the Plains in the 1860s, the Native peoples were chased off their land, the bison were slaughtered, and the new features of the land were the cowboy and the cattle ranch. Cattle ranching was profitable because cows could be transported by rail to the densely populated eastern states. Although the numbers of cattle were still controlled by water and grass in the long term, just as the bison before them, the chance of quick profits led to overgrazing and exposure of bare soil. The problem of accelerated erosion had begun. Worse still, farmers arrived in the Great Plains to *cultivate* the hard, dry soil. Unfortunately, these homesteaders were mostly immigrants from northern Europe and they were accustomed to plentiful precipitation. Their farming methods, their seed, their ideas — all were geared to moist climates. But by practising these techniques in the dry Plains they doomed their land to disaster. Right from the start they had to face the lack of water. In times of severe drought only windmills saved them. With these they could pump sub-artesian water from subsurface rocks, but this was only enough for human use; there was little they could do to solve the water shortage on the fields. These farmers ploughed up the grass and left the soil bare while they waited for more rain to fall or for their seeds to germinate. But with no protection during this vital period, slowly the topsoil began to blow away (Fig. 12.5).

By the 1930s wind erosion was so severe on some days that soil was carried right across America. For example, on the morning of 11 May 1934, the people of Boston stopped in the streets to look at the dirty yellow sky and realized that millions of tonnes of Great Plains topsoil were being blown into the Atlantic Ocean. After episodes such as this it became clear to people throughout America that something had to be done. But wind erosion in the area that became known as the "Dust Bowl" was not the only problem. When the occasional rains did come, they were downpours of such intensity that the soil could not absorb the water. Consequently, the surplus water ran over the surface into the furrows and it carried even more topsoil away.

Figure 12.5 Wind erosion In areas with low precipitation, soil is soon blown away even while it is being cultivated.

SUMMARY: SOIL EROSION BY WIND AND WATER

Wind erosion occurs when a soil is left without a complete vegetation cover. It is most effective with:
(a) strong winds with sufficient energy to pick up soil (usually on plains);
(b) soil particles that are poorly stuck together with organic gums (usually soils that have not been manured);
(c) ploughing parallel to the wind direction which results in wind being funnelled into the furrows and speeding up;
(d) a long period of drought so that the soil can dry out and become dusty.

The areas of greatest risk are plains used for arable cultivation, e.g. the Great Plains and the Prairies of the USA and Canada; the Steppes of Russia; the Sahel region of Africa; the central plains of Australia; and, on a smaller scale, the semi-arid regions of the Kalahari Desert in southern Africa, and the new polders of the Netherlands.

Water erosion occurs when a soil is left without a complete vegetation cover. It is most effective with:

(a) precipitation greater than can be absorbed by the soil (usually in hot regions where heavy thunderstorms are common);
(b) precipitation with large droplets that can smash up soil clods and splatter small soil particles down slope or into soil pores (again most likely with thunderstorms);
(c) steep slopes where water can run quickly over the surface, form small streamlets and erode shallow channels called **rills**. Rills are only a few centimetres deep, but they can merge into bigger channels called **gullies** which can be several metres deep.

The areas of greatest risk from water erosion are those that receive thunderstorms. They include all of the hot regions of the world where the land is under arable cultivation or where the soil has been exposed by over-grazing.

The Sahel region has one of the largest gullied areas in the world.

Soil exhaustion ## Too much of a good thing

The problem of water erosion that helped spell disaster in the Prairies and Plains also plagued the farmers of the southeastern USA. They had a hot climate with plenty of precipitation and very fertile soils. Indeed the soils and climate were so good that high-yielding crops, such as cotton and tobacco, could be grown year after year. However, inevitably the nutrients in the soil were used up by the plants faster than they could be replaced by natural weathering. Slowly the yields went down, but there was plenty more land left, so the farmers just moved on to new, uncultivated land.

Cotton and tobacco grow best in well spaced rows. Between the rows the ground was kept hoed and bare (Fig. 12.6). In these areas, the precipitation occurs in heavy downpours. As it fell on the bare soil — a soil depleted

Figure 12.6 Cotton growing in Nigeria

in humus and nutrients by continuous cropping — it ran over the surface and washed the topsoil away, million tonne after million tonne.

Only after many years of such mistreatment did the farmers realize they had ruined thousands of square kilometres of land.

Understanding soil erosion

Lessons to be learned

It is a hard job trying to repair the destruction of so many years. The core of the trouble in all of these cases of erosion was that the farmers simply did not understand the delicate balance between climate, plants, and soil. It was difficult for them to see what they were doing wrong because it can take years of mismanagement for a soil to deteriorate so badly that it will not produce a crop and become vulnerable to erosion. During this time the soil is losing its humus, its gums, and its nutrients, but for a long time this does not show. Then one year farmers will be faced with a soil blowing away, or a crop with a very low yield, even though they have done nothing special that year. As a result they are left puzzled. Of course we now know that plant leaves protect the soil from winds and plant roots bind it together. We now know that as leaves die they are incorporated into the soil to provide humus, gums, and nutrients. Further, we now understand that humus retains moisture and that the gums stick soil particles together. In this way easy routes for water movement are provided and clumps of soil are created that are too big to be blown away.

With a hundred years of research behind us it seems obvious that if we take the natural plants away in order to grow crops we must somehow put back into the soil the nutrients and organic matter that it needs. But the first research was done by the farmers the hard way — by trial and error, and a willingness to learn.

Soil conservation

Repairing the damage

There were many lessons to be learned by decades of soil erosion and most were learned too late. By the end of the 1930s even winter wheat would not grow in the Great Plains. The soil was deprived of humus, nutrients, and a plant cover, and it had simply dried up.

A soil needs years of tender care to rebuild. In the "Dust Bowl," the topsoil had been blown completely away from some areas. Other places were so badly gullied by water running over the surface that the land could not be worked with machines. The most severely gullied land had to be abandoned, although it was planted with grass and trees whose roots would reduce further erosion. The remaining soil was protected from wind and running water by planting clover or alfalfa. These crops have broad leaves that absorb the force of heavy rain, they have complex trailing roots which bind the soil, and they also produce a surplus of nitrogen. When these crops were ploughed into the soil they added both organic matter and available nitrogen.

However, farmers needed a lot more help than this. They had to be taught to return wheat stubble and straw after harvesting, both to put back some organic matter and also because straw, spread over the soil, helps prevent wind erosion (Fig. 12.7). On the steeper slopes of the southeastern USA they learned to plough with the contours instead of up and down the slope.

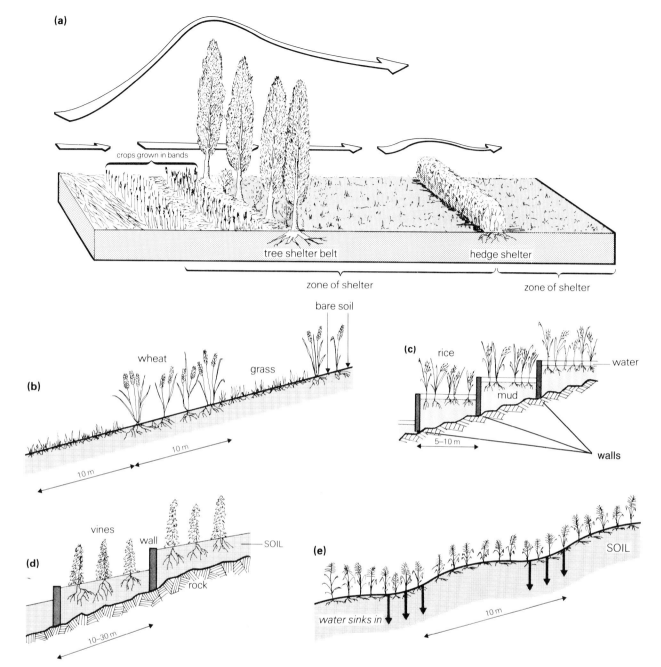

Figure 12.7 Soil conservation methods
(a) Protection from wind using shelter belts; (b) alternate crops across a slope; (c) rice grown in water enclosed by terrace walls (see also Fig. 1.23); (d) vines on terraces (see also Fig. 5.4); (e) slope ploughed to produce soil terraces; (f) contour ploughing

Exercise 12.1 Soil erosion

An experiment was conducted to find out how much soil was being lost by erosion in East Africa. For an experimental site the results were as in this table.

Land use	Soil loss in tonnes/hectare
grain	80
forest	0
bare soil	150
grass under controlled grazing	1
grass being overgrazed	120

1. Write these figures in order of size and land use suffering the least erosion.
2. Plot a bar graph of the data.
3. Explain the lessons to be learned from these results which might be of benefit to the farmers.

This prevented furrows acting as river channels and rapid run-off causing erosion after rainfall.

Cattle ranchers had to learn to control the numbers of animals on their land so that overgrazing was prevented; cotton farmers had to be taught that there were often better crops that could be grown on their land and that fertilizer application was essential. They had to be taught how a layer of rotting vegetation (called a **mulch**) laid over the bare soil between the rows would not only stop erosion but also add humus to the soil. In a little town in Alabama there is even a monument erected to yet another lesson. The monument commemorates the time when an insect called the boll weevil destroyed the cotton crop and forced the farmers to change from **monoculture** (growing only one crop) to growing a variety of crops such as peanuts and fodder crops and keeping dairy cattle. With a wider range of crops these farmers are now more prosperous than they have ever been, because if one crop fails they have others to rely on.

Today most of the Great Plains and the Cotton Belt are once again thriving agricultural regions. Although the farmers suffered badly from mistakes and many had to abandon their land and look for work elsewhere, those that remained were able to call on the resources of a prosperous country and high technology to help them. The farmers were able to obtain bank loans and the government was able to give grants and subsidies until

---

SUMMARY: SOIL CONSERVATION

Soils at risk from erosion must be protected from strong winds and heavy rain. The main methods for preventing wind erosion are:

(a) planting shelter belts of trees across the path of the wind (Fig. 12.7); this is not widely used because trees are effective shelter only for short distances, and they use up space and soil water, can shade crops and thereby prevent crops ripening:
(b) planting crops that ripen at different times of the year in alternate bands across the fields; only small strips of soil are left bare at a time;
(c) spreading straw over the soil as soon as ploughing is over (Fig. 12.7a); this prevents the wind reaching the soil and drying it out, makes the ground surface rough, and slows down the wind;
(d) sowing the seeds of the next crop through the remains of the harvested crop using special machinery; with this method the soil is never laid bare.

The main methods for preventing water erosion are:

(a) sowing the seeds for the next crop through the remains of the harvested crop; no soil is then exposed to beating rain, and even if water runs over the surface the soil will be held in place by plant roots;
(b) ploughing across the slope (contour ploughing) rather than up and down; surplus water then collects in the furrows and slowly soaks away;
(c) planting grass strips across the slope parallel to arable crops (Fig. 12.7b). Any soil lost from the bare soil will be trapped by the grass strips below;
(d) building terraces to reduce the slope of the soil and to encourage water to sink in; terraces can use brick, mud, or stone walls (Fig. 12.7 c & d), or they can be made by ploughing (Fig. 12.7e).

the problems were over. However, today the problems of soil erosion have spread to the developing world, as we shall see later in the chapter. In these areas, although we now know how to combat erosion, there is often not enough money available to tackle the tasks, so erosion increases year by year.

Farming in Florida

Farming in the sun

In the earlier part of this chapter we saw how poor land management caused a loss of fertility and catastrophic erosion in parts of North America. Today in these areas farming methods make full use of the lessons learned. In this part of the chapter we shall look at some examples of artificial ecosystems as they are managed today both in the developed world and the developing world. To make the contrast more realistic we will compare these methods within the tropical and near-tropical world.

Florida — known as the sunshine state — is an example of a region where artificial ecosystems are kept in balance by the use of high technology. Because Florida has a mean annual temperature of over 21°C and high precipitation, (1400 mm per yr) it is well situated to grow citrus fruits such as oranges (Fig. 12.8). Floridian farmers also grow vegetables for the markets in the northern USA, Canada, and Europe during winter when prices are highest. The land, however, is low lying and often swampy. Using large machines the swamps have now been cleared away and kept dry by means of large numbers of ditches. On the cleared land oranges are grown in large numbers. Each tree is uniformly spaced to allow easy access by machinery. Of course, drainage has changed the ecosystem: what was once a dense swamp and water-logged soils with many plants competing for space has become a single species of tree grown in a regimented pattern. Estates of single species crops are called **plantations** and are often run by large companies with access to extensive and distant markets. They do not provide food directly for the local community.

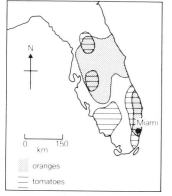

Figure 12.8 Areas of citrus fruit and tomato cultivation in Florida

Because this is an artificial system, farmers have to work hard to control weeds and pests with the widespread use of herbicides and pesticides which are distributed by spray irrigation, wind machines, and aircraft. Occasional frosts are even countered by large oil-burning stoves which are lit in the orange groves when needed.

Without constant attention, vegetables such as tomatoes and cucumbers are even less able to survive environmental stresses than citrus trees. Daily doses of water, fertilizer, and pesticide are provided by overhead sprinklers in a pampered form of agriculture that survives only because of the high prices that can be obtained for the high-quality crops.

Farming in Israel

The green desert

In Chapter 16 there is a description of the normal nomadic way of life of the desert peoples of Morocco. Their survival depends on finding small plant shoots among the pebbles of the desert to provide grazing for their goats. For water they are limited to scooping up from shallow wells using a leather bucket and a long piece of rope. Understandably, with this system the desert will support only a very small number of people.

In the Negev Desert of Israel things are very different. Here water is brought to the area by canal and pipeline so that cereal crops can be grown.

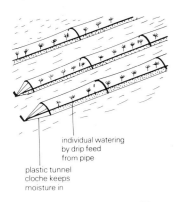

Figure 12.9 Water conservation
Individual treatment is given to plants in countries such as Israel where water is limited.

The animals are sheep, not goats, and they are fattened for market using cereals rather than being allowed to forage in the desert. Citrus fruits and vegetables are also grown, again using irrigation water. Each plant is drip fed from plastic hose pipes. In this totally artificial system a large pumped water supply is the vital means of success. Even so the vegetables are often grown under plastic sheets (called cloches) especially to help reduce moisture loss (Fig. 12.9).

To a limited extent the high-technology desert farming is practised in other parts of the Middle East. For example, in southeastern Libya, some of the oil wealth has been used to change the Kufra Oasis from a region of marginal subsistence agriculture to one where a large surplus can be grown. In this case the vital water is pumped out of a sub-artesian supply 200 m below the surface and used to irrigate fields, so that clover, wheat and barley now grow within sight of sand dunes.

Coffee production

Cold coffee costs more

In the previous examples much *money* and *energy* have been used to grow very *special* crops. However, only small amounts of food can be grown in this way. Most of the world's food has to be grown without such excessive attention. As a result, crops usually have to be grown within the limits of their tolerance to drought, cold, or heat. Nevertheless, there are great pressures on people to try to grow more and more of a crop if it sells well on the world market. To do this they have to grow the crop in more and more marginal regions. Coffee growing is a very good example of some of the results of such expansion.

Coffee comes from a tree which first grew naturally in Ethiopia. It is a very adaptable tree and it will grow in a wide range of tropical and subtropical climates and soils. Gradually, as demand rose, it was introduced into India, Indonesia, and then South and Central America (Fig. 12.10). This was relatively easy because its simple cultivation is suited to less-developed agricultural methods as well as large plantations. Above all, coffee needs little attention and it produces a higher return in money than any alternative crops grown in the fields.

Frost is the biggest enemy of the coffee tree. But despite this limitation, high value on world markets has meant that coffee trees have been planted further and further polewards and at higher altitudes. Similarly, coffee growing has been extended equatorwards (e.g. Sumatra) and to lower altitudes (e.g. into the Caribbean islands) where the climate is really too damp and the trees become very prone to disease.

Figure 12.10 The world's major coffee-producing areas

Figure 12.11 A palm oil plantation in Malaysia

Exercise 12.2 Plantation farming

Figure 12.11 shows part of a palm oil plantation in Malaysia.
1. Explain the advantages of a plantation system for farming.
2. Rows of small trees grow between the mature palms. Suggest a reason for this practice.
3. Use your knowledge of tropical climates and vegetation to list at least 3 costs and 3 benefits of a palm oil plantation.

Not surprisingly, therefore, the marginal areas at risk from frost have increased greatly in recent years. When a severe frost occurs it can destroy much of the crop. This in turn causes hardship to the growers and a rapid rise in the price of coffee on the world market. Something like a fifth of the coffee crop is now frequently damaged by frost, but the continuing demand from the developed countries will mean that the marginal areas are still kept in use.

Farming in the equatorial rain forest

Appropriate technology

The trans-Amazonian highway (BR-364) strikes like an arrow through the heart of the Brazilian rain forest. The road has not been blasted out of solid rock, but bulldozed through the deep red soil that has supported a luxuriant forest for millions of years.

You will remember from Chapter 11 that everything recycles quickly in the rain forest. Teaming multitudes of insects continually gather every fragment of dead organic matter. If you were to leave a newspaper on the ground it might be cut up into tiny pieces and carried away within a couple of hours. Wooden furniture will even be eaten where it stands — from the legs up!

The Native people of the Amazon understand that the rain forest and its environment has a very delicate balance. They also know that although recycling is rapid and the vegetation grows well, the soil is rarely fertile. For this reason, the Native people cut and burn only temporary small patches (called gardens) out of the forest. Native farmers realize that any bare soil is liable to erosion and leaching of nutrients, therefore, the charred stumps of trees remain where they fall and plots are rarely weeded. It is better to keep the soil and its nutrients and tolerate some competition from weeds (Fig. 12.2 b & c).

Exercise 12.3 A rubber estate

1. The map and section (Fig. 12.12) show a rubber plantation set among hills in Malaysia. Why does the estate occupy the area indicated?
2. It is intended to expand the estate into undeveloped areas. Which area do you think would be exploited first?
3. What erosion control measures would be needed if prolonged cultivation of rubber trees is to be successful?
4. If the main estate road were driven up into the hills of area B, how would this encourage erosion?

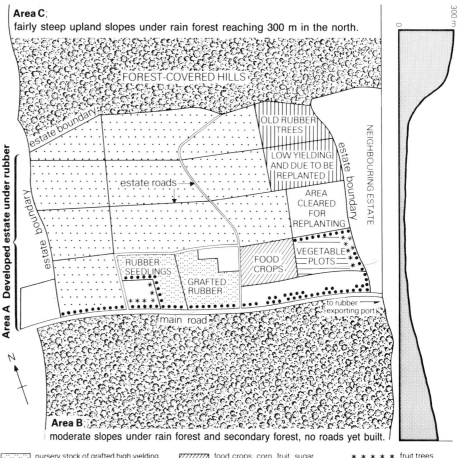

Figure 12.12 A rubber estate

Even with these conservation methods, the soil will sustain crops for only two years before it is exhausted and a new area has to be cut and burned. Even then rain forest soil will produce only enough for subsistence living. But the *Native people have adapted their methods to the environment.* Because of its small population, the vast Amazon will provide food indefinitely – if a blend of gardening and hunting is used.

Very few people, other than Natives of the rain forest, understand the soil's low fertility. At the end of the 19th century, the city of Belem grew rapidly on the banks of the River Para in Brazil. Many people from the drought-stricken northeastern coast arrived, cut down large tracts of forest, and planted corn, beans, rice, and tobacco.

Soon they found out how fragile the land really was. The soil was exhausted within a couple of years and they had to move to new areas...

and new areas... and new areas, until they were forced to return to the original cleared lands which had still not recovered. But the settlers had ruined 30 000 km² in the process. Today there is no longer enough food grown on these impoverished soils to support the 800 000 people of Belem and the government is trying to evacuate them to new areas.

More recently in the southern rain forest, an area called Mato Grosso has been cleared and grass has been planted. Over 10 million beef cattle graze on these lands, but not for much longer – soil fertility is already low and erosion has begun. But still the lesson has not been learned. Now the government is encouraging people to move to the newly cleared areas beside the BR-364 highway. The talk is again of the old mistaken ideas of "fertile valleys" and a "thick layer of humus accumulated during centuries." So far, more than 500 000 km² of trees (more than half the size of BC) have been cut down and burned. The new farmers are going to be the latest in a long line who have all discovered that the fertility of the rain forest soils cannot be maintained without a ruinously expensive use of fertilizers. Meanwhile, the forest devastation goes on...

Desertification

Do people create deserts?

Fourteen percent (630 million) of the world's population live in arid or semi-arid areas. One of the main problem regions in the semi-arid zone is called the Sahel (Fig. 6.11). Sahel means "the border" in Arabic and it is the name used to describe the semi-arid grass, bush, and occasional tree vegetation that forms the dry scrub land zone next to the Sahara desert. Here people have followed a yearly pattern of activity controlled by the seasons. The longer season is hot and almost completely dry, the shorter is hot and humid with rain falling in heavy showers. However, being on the edge of the desert, the precipitation is not reliable and sometimes the land may not receive the 400 mm or so of precipitation needed for cultivation.

Traditionally people from this region were herders of cattle, sheep, goats, and camels (Fig. 12.13). They were nomadic and they moved south in the dry season and then returned north when the rain came. Others cultivated

Figure 12.13 Cattle herding in the Sahel, Niger, Africa
Water and grass are the keys to survival in this changeable region.

Figure 12.14 Gully erosion in Africa following overgrazing

the land in small patches which they clawed from the savanna by burning. When the fertility of the land was lost they burned a new small area and left the used area to recover naturally. In the areas of Central Africa with the poorest soils, fallow could take up to 35 years to revive. Six to ten years is needed even in the better regions, but such long resting periods did not matter as there was plenty of land to go round.

Now the population of Africa is rising quickly. Without sophisticated farming methods to keep pace, the population is exceeding the productivity of the land. Land that was left in fallow for long periods is now used every two to three years, but it does not recover in such a short time and so its yield is very low. The productivity of the fields has actually gone down under this pressure despite the hard work of the farmers.

The herders have fared just as badly. National boundaries have stopped most of their nomadic wandering in search of good grazing. In the Sudan, for example, 3-4 km² of savanna are needed to support a family herd of 30-40 cattle. In 1945 the stock numbers were just equal to the land available.

Figure 12.15 Village of Timia Air Mountains, Nigeria
The difficulties of applying modern agricultural techniques in an environment which lacks adequate soil and vegetation are apparent. Extreme isolation makes development even more difficult.

Since then the human and animal populations have doubled. It is no surprise to find that overgrazing is widespread and the soil left open to erosion (Fig. 12.14).

The clear answer to this problem is in the use of fertilizers and farming methods that use soil moisture more efficiently. The problem is that the people do not understand how to use fertilizer, they have never had any surplus of money to buy it and the isolation of many areas makes for supply difficulties (Fig. 12.15).

At the moment because of overuse of land in the Sahel, soil erosion is turning more and more land into an unusable desert. A new word has been coined especially to describe this disastrous situation — **desertification.** Already desertification has claimed more than 30 million km² of the earth's agricultural land, not because of any change in climate but solely due to people's destruction of the natural vegetation and their inability to stop the resulting soil erosion. It is not that people are being intentionally careless with the land, but desperation for food makes long-term goals difficult to see and accomplish.

We finish this chapter on a note of despair. We now understand a lot about how the environment works and the results of changing the naturally balanced ecosystem. Most agricultural advisers now understand correct management methods. It is vital that this understanding is used, but so far little has been achieved.

Figure 12.16 Future land-use possibilities
(See page 156)

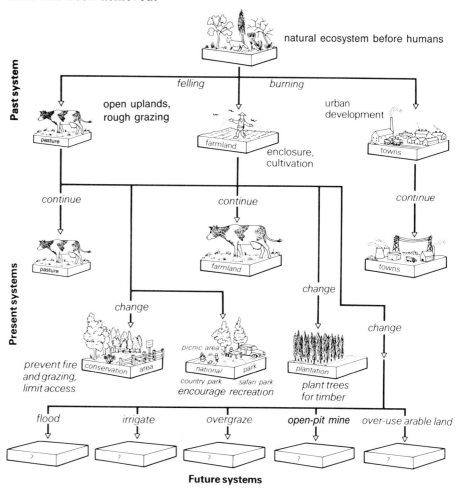

Exercise 12.4 The future of our land use

People have many alternative ways in which they can use the land. Some are destructive, others creative. Some of the choices are shown in Figure 12.16. Five boxes have been left unlabelled. For each box, describe and explain the result of the land use shown.

QUESTIONS

Multiple choice

1. Frost shattering is a process of physical weathering which is most effective on exposed rock faces when
   (a) temperatures are too low for chemical reactions to occur
   (b) no water is available
   (c) temperatures fluctuate above and below freezing
   (d) rocks are not cracked or jointed
2. In deserts where daily temperature ranges are extreme, rocks show almost no signs of weathering unless
   (a) salts are leached from them
   (b) the rock is, like granite, homogeneous
   (c) high temperatures are accompanied by high winds
   (d) water is naturally available
3. The smaller the surface area of a mass of rock, the
   (a) faster that rock is likely to weather
   (b) slower that rock is likely to decompose
   (c) more prone to erosion that rock is likely to be
   (d) more likely that rock is to experience mass wasting
4. In the C (bedrock) horizon of a soil, almost all weathering is due to
   (a) chemical decomposition
   (b) daily freeze-thaw
   (c) physical disintegration
   (d) mass wasting
5. A soil which would most likely experience little or no benefit from the addition of lime is known as a
   (a) podzol
   (b) chernozem
   (c) bog soil
   (d) latosol
6. Geographers are expected to explain the present distributions of people, plants, and animals. Therefore, they need to
   (a) understand the evolution of life and landscapes
   (b) remember where all the features of the world are located
   (c) be able to predict hazards such as earthquakes and avalanches
   (d) be able to map the migrations of continents and ocean islands
7. Ecosystems provide geographers with frameworks for studying
   (a) international trade in agricultural products
   (b) the agricultural productivity of new farmland
   (c) the causes and effects of all disasters
   (d) communities of plants and animals and their soils and climates
8. The process by which flourishing plants change their own environments to their detriment leads to
   (a) evolution
   (b) survival of the fittest
   (c) plant succession
   (d) environmental decay
9. When soils are deprived of most of their nutrients and natural structure they become unproductive and
   (a) require ploughing and terracing
   (b) can only be expected to improve
   (c) must be used for industrial purposes
   (d) are prone to accelerated soil erosion
10. Desertification has removed more than 30 million km^2 of land from agricultural productivity because of
    (a) people destroying the natural vegetation
    (b) the warming of the earth's atmosphere
    (c) the destruction of the ozone layer
    (d) rapid destruction of tropical forests

Short answer

1. Explain how a soil forms. Illustrate your answer with labelled diagrams.
2. Explain the terms texture, humus, and soil profile.
3. Name an example of an area with (a) tropical rain forest, (b) tropical grassland (savanna), (c) temperate grassland (prairie or steppe), and (d) tundra. Explain how people have changed the pattern of the natural vegetation in the tropical rain forest.
4. Describe some of the methods used to control soil erosion.
5. Name and locate one example of small-scale subsistence farming. Describe the methods used and the crops produced. Explain how these are influenced by physical, social, and economic conditions.
6. Explain why a tropical rain forest has the greatest variety of trees and the deepest soils of any place on earth.
7. Referring to specific examples and named areas, show how natural vegetation is affected by climate, soils, and people.
8. List at least five tropical crops which are often grown on plantations. Note two features these crops have in common.
9. Draw a map of Africa. Colour the Sahel and identify the region's northern and southern boundaries. List three major problems experienced by the people of the Sahel.

Part four: Shaping the land

Moving water, ice, and wind continually remove soils and rock, exposing new rock to weathering, and changing the landscape. They also scratch and tear at the rock, producing sets of distinctive landforms.

Figure 13.0 Landslide! The roads at the top and in the middle indicate the scale. The photograph was taken in the foothills of the Himalayan Mountains, Nepal.

13 Mass movement processes

Soil movement

On the move

Soil is the result of rock weathering. It appears to be an unchanging part of the landscape because we rarely see it move. We tend to think of a landslide as something unusual and unnatural. In fact, except on flat land, soil is always on the move, creeping downhill a few millimetres a year — an unnoticed part of landscape sculpturing. The soil appears unchanging because new rock weathers to replace that being moved away and the change is so slow that there is plenty of time for well-developed soil layers to form.

The steeper the slope, the faster soil moves downhill. You may have noticed that steep slopes have thin soils. This is because weathering of new rock can only just keep pace with movement. Another sign of rapid movement is that soils on slopes contain many pebbles of unweathered rock. Sometimes boulders move so often that chemical weathering never has any real chance to occur. In these cases, movement is faster than soil formation and a slope of boulders is produced.

Scree formation

Block by block

On very steep slopes, where soils do not have the time to form, bare rock is exposed and mechanical weathering becomes important. Blocks are slowly wedged loose and they fall, bouncing and rolling until they come to rest. Mountaineers on frosty rocks are well aware of the importance of frost shatter. As the morning sun warms the rocks, falling blocks broken from the rock face overnight can be a major hazard. These blocks are mostly large and sharp-edged and so do not travel far. They soon build up as piles of loose rock called **screes** (Fig. 13.1).

Figure 13.1 Scree slopes in Lillooet area, southern interior of British Columbia

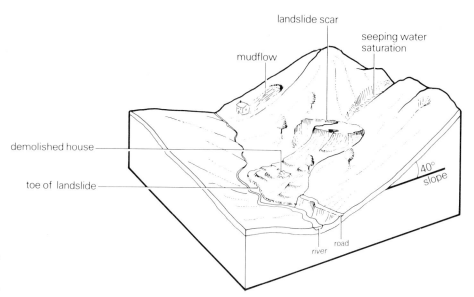

Figure 13.2 The events of South Topanga Canyon, Los Angeles, USA

Landslides

Slipping and sliding

When slopes are less steep, screes do not form and weathered blocks remain in place, eventually making a soil. Nevertheless soil can be very unstable.

We often hear of landslides in the news media, especially when houses are destroyed or people hurt. In South Topanga Canyon near Los Angeles (California), for example, a man was awakened at 3 am one morning by a loud rumbling sound outside. He later told local journalists that just as he opened his eyes the house collapsed and the next thing he knew was that he was in hospital. A neighbour was also awakened by the noise only to find a mudflow oozing in through a back window (Fig. 13.2)!

What had happened to allow gravity to work so rapidly on the slopes in California? It was as though the weight of soil had been too great for the angle of the valley slope and the whole mass had simply slid downwards. Yet nearby slopes of similar angle had not failed. This is because there is a fragile balance on all slopes between the mass of the soil and the amount of friction opposing movement. The presence of large amounts of water in a soil is one very good way of tipping this balance towards disaster.

The landslide and mudflow in South Topanga Canyon occurred partly because a lot of precipitation had fallen the day before. Water soaked into the soil and caused some of it to become saturated. The extra water added great mass to the rain-soaked soil and increased the fluid pressure where the soil was in contact with the bedrock. This made the soil and weathered rock unstable. Eventually the friction between rock and soil was broken; soil broke loose in liquefied slabs and slid quickly down the valley into the houses below.

Soil landslides sometimes turn into mudflows, depending on the amount of water present and the type of soil. Sandy soils tend to flow; clay soils tend to slide or slump as slabs.

Landslides mostly occur when soil on steep slopes becomes saturated. This does not happen very often and most of the time water can drain out

Figure 13.3 Landslide sites
This meandering river is undercutting a slope at the bends, causing landslides to occur.

of the soil as fast as it soaks in. So it takes a very long period of heavy precipitation to cause water to build up in a soil and allow a landslide to occur. Landslides also occur if valley slopes are made steeper, such as when they are undercut by a river (Fig. 13.3).

Although landslides occur only occasionally, they are always dramatic. Landslides near rivers often dam up water temporarily. They are also probably the most important way in which soil moves rapidly on steep slopes.

Figure 13.4 Terracettes on a steep coastal slope
The sedimentary rocks of this rugged coast have weathered into terraces. Where soils have formed and been covered by grasses on the steep upper slopes minor slumping has formed many terracettes.

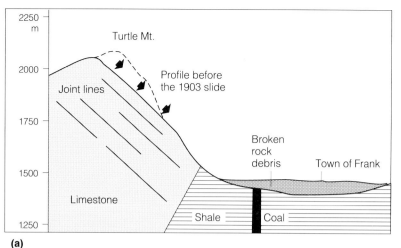

Figure 13.5 A massive landslide at Frank, Crowsnest Pass, near the BC-Alberta border

Exercise 13.1 Landslides

1. Figure 13.5 shows part of a massive landslide at Frank, Alberta. Explain what has happened.
2. Locate Frank and Crowsnest Pass in an atlas. Draw a sketch map of its location and name the mountain range. Add a copy of the cross section to your answer.
3. Disaster struck Frank at 4:10 am on 29 April 1903 when 90 million t of limestone crashed down from Turtle Mountain, buried a coal mine, the railway, and part of the town, and killed 70 people in about 100 seconds. List three things that people might have done to minimize the damage caused by the landslide.

Figure 13.6 The extent of the mudflow which developed from a rock and ice flow in the Andes of South America

Catastrophe! *Severe landslides*

Very occasionally massive landslides occur, involving rock as well as soil. In these rare cases rock and soil can move at speeds up to 100 km/h and cause disasters. There are many reasons for large landslides; they are significant because they may move as much material in a few minutes as a river can move in a thousand years. It is also important to realize that one type of movement can often trigger off a whole series of others.

The following example describes the largest single movement of debris in recent times. It started when an earthquake caused the collapse of part of the summit ice cap of Mount Huascaran in the Peruvian Andes of South America. As the ice cap fell away it dislodged thousands of tonnes of rock which, together with ice, went hurtling down a nearby valley. The valleys of the Andes are so steep that the mass of rock and ice, instead of coming to rest, gathered speed, and as it burst into the valley below it picked up thousands more tonnes of rock, water, and soil (Fig. 13.6). Some of the pieces of rock being carried were bigger than a house! As it went down the valley, the ice in among the rock fragments melted and the mass of material turned into a mudflow. Fifteen kilometres down the valley, the people of Yungay heard a distant roar. One of the few eyewitnesses to survive the next few minutes described the scene:

> The crest of the wave [of mud] had a curl, like a huge breaker coming in from the ocean. I estimated the wave to be at least 80 m high. I observed hundreds of people running in all directions [and] all the while there was a continuous loud roar and rumble. I reached... the cemetery near the top just as the mudflow struck the base of the hill and I was probably only ten seconds ahead of it. It was the most horrible thing I have ever experienced and I will never forget it.

Soil creep

But quietly creep the hills

We have started our description of hillslope processes by looking at the most spectacular. They are important processes because, although rare, they cause much movement. But there probably isn't a landslide or an avalanche of rock and ice outside your home and you are unlikely to live on a steep slope, so what is happening everywhere else?

Here we meet one of the difficulties of landscape study. Some of the processes are spectacular and get all the news; others work all the time but in such small steps that only delicate instruments will detect that they are even working at all. The most important slow process is called **soil creep.**

Most soil creeps downhill very slightly every time precipitation occurs. This is because clay particles swell as they get wet and shrink again as they dry out. During this change of size the soil is being pulled downslope by gravity and the swelling and shrinking with each storm help to ease the soil downslope. This is an extraordinarily slow process and at the end of a year a piece of soil might have moved only 1 mm downslope. Soil creep works fastest at the surface where wetting and drying occur most often. As a result, sometimes the process can be seen indirectly. A telephone pole or a wall may lean over as soil creep pushes against it. Even trees may be pushed over (Fig. 13.7). As they continue to grow, their trunks bend to retain an upright position. During the winter, freezing and thawing of water in the soil also causes swelling and shrinking and this adds to soil creep. Thus soil creep may move a piece of soil perhaps only 10 cm down a slope in

Figure 13.7 The apparent effects of soil creep

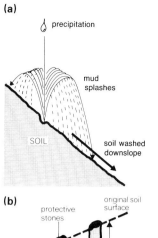

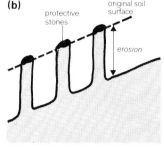

Figure 13.8 Surface wash
(a) Heavy beating precipitation splashes the soil away. (b & c (*right*)) If large pebbles protect the surface, then earth pillars will survive for a while. These earth pillars are part of the Hoodoos in southeastern BC.

100 years; hardly worth considering, you might think. But every particle on the soil surface moves by similar amounts all over the hillside. So if you add all of these movements up they can be as important as a localized landslide. Scientists have still not made enough measurements to be sure of the answer to this, but we do know that landslides can work only on steep slopes; thus, soil creep is possibly the most important way of moving soil particles on gentle slopes. On the other hand, on steep slopes soil creep is probably less important than landslides.

We tend to take grass and other vegetation cover for granted, even on steep slopes. The roots of plants bind themselves and the soil together and so reduce soil movement. Hillslopes change very slowly under these conditions. However, if we lived in a low precipitation or low temperature area we would not be so used to vegetation. In these areas both vegetation and soil are uncommon. On steep slopes in well jointed rocks, screes occur, whereas weak rocks are extensively cut into by channels — called **gullies** (Fig 13.8d). These features are common any place where water can flow over the surface. They are especially common in semi-arid regions, but where people have cleared land in the tropics the effects of heavy precipitation are also clear to see (Fig. 13.8a-d).

(d) Eventually the land will become heavily gullied.

Soil solution

Before your very eyes

In the previous sections we have looked at the movements of soil particles. However, there is also a very important way in which soil material moves completely invisibly. We are referring to that part of the soil taken away in solution. We should not be surprised to discover that solution is important because we know that soil is formed by chemical weathering. Soluble materials remain after rock has been chemically weathered to clay, and they are easily transported away in the soil water. Even less noticeable than soil creep, solution, and consequently the removal of rock constituents (**leaching**), is happening all of the time. Slowly but surely, on steep slopes, on gentle slopes, and even on flat ground, the landscape is dissolving away. The ground is dissolving away beneath you right now!

Of all the hillslope processes, leaching is probably the most difficult to understand and measure. There is much work still to be done, but scientists now think leaching may sometimes account for half of all the material lost from a hillslope in areas which have adequate precipitation. So *leaching can be as important as soil creep, landslides, and gullying all put together* — and yet we can never see it at work.

One final point is worth emphasizing. It is a popular belief that only limestone rock is affected by chemical action, but other rocks also are subject to chemical attack. The rate of chemical decomposition varies widely with the type of rock and the climate. This means that *all* landscapes are affected by solutional erosion.

The nature of avalanches

Avalanche!

Cold regions experience types of movement that are not found anywhere else. Taken together they make up a very special sequence which is described below.

In Antarctica there are some places where temperatures remain below freezing throughout the year. Here frost shatter does not break off pieces of rock and the landscape ceases to evolve. In most northern cold regions, however, spring arrives by about June or July and lasts only a short time. There is no "summer" and by September conditions are wintry again. Now although spring, with its temperatures creeping above freezing, does not last long, thawing allows a whole range of activities to occur quickly. Earlier in this chapter we mentioned how blocks of rock can be dislodged by freeze/thaw processes to produce scree slopes. But on slopes of less than about 40°, loosened debris tends to remain in place. At the same time the slope becomes gentle enough for snow to accumulate, allowing a new process to take over: we all know its name well — **avalanche!**

Winter snow can be the delicate snowflakes that clothe mountain tops like the icing on a cake. It can also provide the material that can come crashing down on a town and blow its houses to bits as effectively as a bomb.

Throughout winter, snow falls and settles on mountain slopes. Because it is light it is whisked up by the winds — cleared from some places but piled up in others. And this is where the trouble starts. A few metres' depth of snow clothing a mountainside may weigh 1 million t. On some slopes this mass builds up an intolerable stress, the delicately interlocking snowflakes are torn apart, and avalanches begin.

For a moment consider the way snow accumulates on a steep mountain slope in winter. Every few days a few centimetres or more of snow settle or are blown onto the slope. It is cold. A snowy day may be followed by a clear freezing night. The snow settles and compacts; the snowflakes freeze together more tightly, lose their angular form and become more rounded. It snows again and again building up a thick layer of snow, which is powdery and made of finely shaped crystals at the surface, and more compact with old, more rounded flakes below. Some of these crystals deep in the snow eventually behave almost like ball bearings, enabling the whole snow pack to move more easily. In the spring, snow and mountain rock are warmed by the sun. But rock and snow warm at different rates and, as the rock faces that surround the snow warm up, so they gently heat the snow from below. By the spring only a very small amount of heating may be all that is needed to cause instability.

Whether it be from a heavy snowfall or spring thaw, the start of an avalanche can hardly be detected: in a small patch of snow perhaps just a few metres across, delicate snowflakes tear apart and allow the snow to slump forward onto the untouched slope below. With the whole snow mass delicately balanced, the new mass causes the snow on this part of the slope to slip forward as well. Soon a chain of action and reaction is started which grows every second. What was only a small slip ten seconds earlier is now a plunging mass 100 m across (Fig. 13.9).

At the start, the slipping snow moves slowly — a few metres a minute. But as it grows bigger, so its speed increases until, half a minute from the start of the avalanche, the bulk of the snow is moving at 80 km/h. By now nobody can outrun it.

As the mass gathers speed it shakes off more and more powdery surface snow into the air. This powder becomes mixed with the air to form a high density "gas," ten times as dense as air. This, too, is pulled downslope by gravity, but is not subject to the frictional resistance of the slipping mass. As a result it starts to move ahead of the main avalanche, picking up more and more powdery snow so that it is soon hurtling down at over 150 km/h and dragging in eddies of air whose speed exceeds 300 km/h.

At this stage the powder avalanche hides the main sliding avalanche in a bubbling mist. The powder is now moving so fast that the air in front does not have time to get out of the way. Pushed downslope by the powder avalanche, the air becomes a compressed shock wave with a devastating power of its own.

In the valley below, the small plume of white snow first seen high on the slope becomes a long column of bubbling whiteness whose approach can be heard as a low rumbling note, vibrating the rock of the mountain. Less than a minute after the first snowflakes have torn apart and caused that slip on the mountain, a million tonnes of snow and a hundred thousand tonnes of air may have hit the valley bottom at over 300 km/h.

For buildings in the path of the avalanche, there is little hope unless they have been specially built. But the importance of an avalanche for landscape formation is that it tears away frost-shattered rock from slopes, and hurtles boulders down to the valley floor. Thus it is a major source of erosion in mountains and it must, in the past, have propelled millions of tonnes of rock fragments onto the surfaces of glaciers to add to their load of debris.

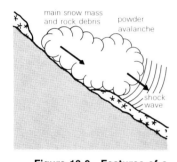

Figure 13.9 Features of a snow avalanche

Building on steep slopes

A recipe for disaster

We can now move on to discuss the way in which soil movements affect our activities. No movement can be helpful to a road, a house, or a dam. In some cases it might just be possible to put up with the effects of soil creep, but even this is unlikely. Clearly, landslides cannot be tolerated at any time. It is therefore vital not to build in an area of slope instability. Planners and engineers cannot afford to ignore the lessons of the Frank slide because many lives could be lost if a mistake were made.

Hong Kong regularly receives large amounts of precipitation. So in June 1972, when 200 mm fell in one day, there was little surprise. However, this deluge was followed by similar amounts on both of the following days. In total the precipitation that fell was equivalent to a whole year's precipitation at Victoria, BC. Towards the end of this rainy period, a giant landslide started to form on a steep slope above one of Hong Kong's most densely populated high-rise areas. Within a space of just five minutes the slide had swept across two roads and completely demolished a twelve-storey block of apartments. Within those same five minutes, sixty-seven people living in the apartments had been killed.

Now from our hillslope studies, we can see why this happened. In humid tropical regions, chemical weathering quickly rots solid rock to great depths, so deep soils cover even steep slopes. We also know that steep slopes and deep soil are an unstable combination. We can also see how people turned an unstable situation into a recipe for disaster. The trouble stemmed from the large population of the colony and the small area of available land. The gently sloping sites were occupied long ago and people were forced to turn to areas of steep slopes. As a result, we have a conflict between human pressure for building sites and dangers from unstable land. The results are shown in Figure 13.10.

Figure 13.10 Slope hazards created by people
(a) This slope is stable because the vegetation helps to anchor the soil to the rock. (b) This slope is less stable. The cutting made for the road gives a short, steep, unprotected slope above the road. Part of the road is built on "fill" (soil and rock dug out of the cutting). This material is very liable to being washed away. (c) This slope is highly unstable. Although the building is securely anchored to the underlying rock, the soil at the top of the slope is unsupported and liable to slip — for this reason, a retaining wall often has to be constructed.

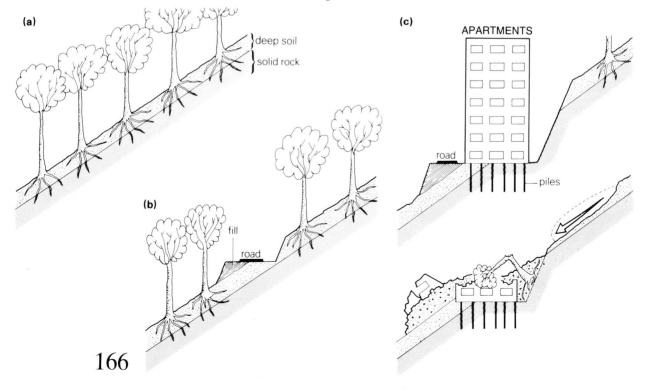

166

Finally, you will remember that water moving through a soil does not always travel in straight lines. Sometimes it can converge, even on a uniform slope, and when it does, the soil gets ever closer to saturation. Above that Hong Kong apartment block, water converged and saturation was reached on 16 June 1972.

Building on frozen ground

This seemed a firm foundation

Cold conditions affect not only mountainous areas covered by ice. Many areas of the plains of Siberia and the Canadian Arctic are extremely cold, but they do not receive enough snowfall each year to compact into ice. Nevertheless these areas are underlain by frozen ground called **permafrost**. Soil and rock, and the water contained within them, are all frozen by the long, cold winters down to depths of hundreds of metres. But the frozen ground is very fragile: warm it ever so slightly above 0°C and the ice turns back into water and the top metre or so of ground melts, turning the soil into a sort of sludge. Normally this is of little consequence to people because such regions have traditionally been sparsely populated. However, in recent years the discovery of minerals — especially oil — under these frozen wastes has changed all that. Now there is an urgent need to understand the way permafrost behaves because vast unspoilt lands and hundreds of millions of dollars depend on it.

For example, about half of Canada and all of Alaska are underlain by permafrost. The discovery of the Alaskan oil field and the need to build a 1300 km-long pipeline, oil rigs, drilling wells, storage tanks, and homes has recently posed a whole variety of new problems. The trouble comes from trying to build on the "skin" of the permafrost that turns into a waterlogged sludge as soon as it is heated. Thus a pipeline carrying hot oil would melt the surrounding permafrost. With the solid support removed,

---

SUMMARY: TRANSPORT DOWN SLOPES

Soil and rock debris move down a hillside or cliff because they become unstable. There are many forms of instability: very fast sudden movements that we call rock falls, landslides, or mudflows; and more gentle, long-term movements called soil creep, slope wash, or solution.

On very steep slopes, soil or vegetation are missing and the main processes at work are mechanical — blocks are broken away by heat or cold (exfoliation and frost shatter) giving **rock falls** which accumulate at the slope bottom as a **scree** (Fig. 13.1). Screes are not usually very thick but they can cover slopes and prevent further mechanical weathering. Unless streams continually wash the scree away from the bottom of a slope the debris will gradually rot where it lies and form a soil.

Moderate slopes are too gentle for scree material to move and so they have developed soils. The movement of soil is different from scree because the particles are small and can be moved with the aid of water. **Landslides** and **mudflows** are typical fast soil movements on slopes. After heavy precipitation, water may build up in the soil and help to reduce the pressure of one particle on another. In clay soils a whole slab of material slides downwards, leaving a crescent-shaped scar behind. When the soil is sandy, water build-up often causes the soil to flow downhill. Where soils are very thin, soil slumps in strips creating **terracettes.**

On more gentle slopes **soil creep, solution, rainsplash,** and **surface wash** are important (Fig. 13.8). Soil creep results from swelling of soils as they are wetted by precipitation, and shrinking as they dry out again. It can also be produced by freezing and thawing of water in the soil pores. Surface wash occurs with very heavy precipitation in the semi-arid areas where there is not proper vegetation cover to hold the soil in place. It often produces **gullies.**

Finally, although we can never see solution and **leaching** at work, we know that it is the main way of forming a soil and so must be important on most slopes. In lowland areas, leaching may remove as much material as all the other movement processes combined.

the pipe would settle irregularly into the sludge and it is likely that somewhere along its 1300 km it would crack (Fig. 13.11). The oil spillage that could result is too serious to risk. Therefore, the pipe now runs aboveground in an insulated jacket.

In the past, oil storage tanks have been built with their foundations in permafrost, but over a long period of time heat from the oil has melted the permafrost below the concrete support raft, so that the tanks have settled and are prone to rupture.

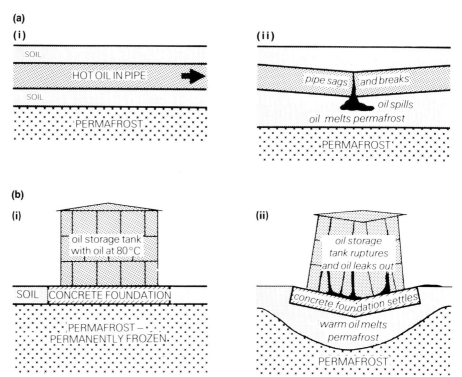

Figure 13.11 Permafrost hazards

Exercise 13.2 Problems of waste disposal

Waste comes in a variety of forms. Industrial waste can be solids such as slag, rock, or ash from a furnace; it can be liquid such as toxic chemicals used to clean metals; it can even be the radioactive waste of a nuclear power station. Then there is domestic waste — plastic bottles, old appliances, cars, paper, etc.

The easy answer is to fill up ravines or dump the waste on unused ground. But is it the right answer?

1. In Figure 13.12a you see two pits that will be filled with solid waste. On the surface they look the same, but underground conditions are different. Explain why one of the pits is safe and the other one is not safe.
2. Now consider highly toxic chemical waste. Are any sites safe?
3. In Figure 13.12b you see two piles of mine waste which have been dumped on surfaces which looked the same. Again underground conditions are different. Explain why pile (i) is safe but pile (ii) should be removed immediately.
4. Figure 13.12c shows a cross section of the town of Aberfan, Wales, where coal waste was dumped on a slope above the town. At 9:15 am on 21 October 1966, 140 000 t of waste flowed down the hill killing 144 people, 116 of whom were schoolchildren. Why was this an unsatisfactory site?
5. What problems are associated with "sanitary landfill" in your neighbourhood? What alternatives are being explored?

Figure 13.12 Hazards (a) Garbage dumps; (b) mine spoil dumps; (c) coal spoil dump at Aberfan.

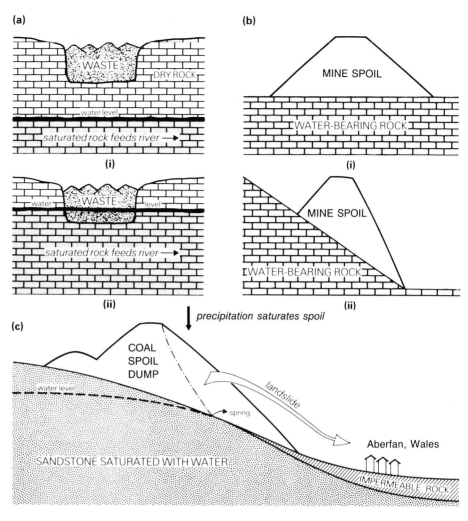

Exercise 13.3 Road building in Nepal

Nepal is a developing country with a very mountainous landscape. Aid programs are helping to provide better links through the country and with neighbouring India, but there are severe problems created by deep river valleys. Roads connecting towns must have numerous hairpin bends and bridges, all built at an altitude of more than 3000 m above sea level.

1. List the types of movement you could expect to find in an area of steep slopes. A preliminary survey of the line of a road was made from maps and air photos, but the dense vegetation hid most evidence of slope movements. A ground survey found slope movements in the places shown in Figure 13.13.
2. Why should slope movements be most common in the region labelled A?
3. Field surveys made engineers reconsider their original route line. List the major problems the route would have encountered.
4. Suggest an alternative route for the road. Explain how your route is better than that in the original plan.

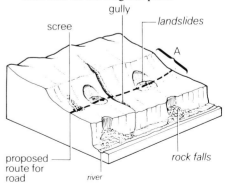

Figure 13.13 Slope movements found from a ground survey in Nepal

169

14 Rivers and valleys in humid regions

The water cycle and landscape formation

What goes up...

Until the first rockets were fired into deep space, one of the most fundamental laws of nature was "what goes up must come down." Thus, although water is lifted as vapour from the oceans using the sun's energy and carried high into the sky where it forms clouds, it must eventually fall back to earth as precipitation.

Cloud droplets are very small and light but they are still pulled to earth by gravity. As they grow, they become so heavy that even updrafts of air cannot keep them aloft and they "fall" as rain. But the pull of gravity does not stop when precipitation reaches the ground: water is slowly pulled through the soil until it reaches a river bank. Having reached a river, it is finally pulled quickly down to the sea.

If we look at the water cycle in this way — an initial upward movement using heat energy from the sun followed by a long downward return using gravitational energy — we see that water on land is really pulled through soils and rocks and along river channels. Any winding river channels, small soil pores, and so on, are merely inconvenient obstacles thwarting the return of water to the sea. Anything the water may do to pieces of soil or rock, such as help in weathering and transport, is purely incidental. Erosion is just a normal by-product.

Internal loss of energy

Water at work

Most of the energy available to a river is used up simply in making water flow. This becomes apparent sometimes when water flows quickly in a pipe. In hydroelectric power stations for example, when water is taken from a high-level reservoir to a generating station in a valley bottom, the transfer pipes become warm, as friction between the water and the sides of the pipe is converted into heat. Over 95% of all energy in a stream may be used up in overcoming friction, which leaves relatively little energy to transport debris or erode bed and banks.

As water rumbles along in its channel much energy is used in pushing against the pieces of rock that form bed and banks. If a piece of rock is pushed hard enough by water flowing past it, then it too will begin to move. Sometimes these pieces are already loose on the river bed and are transported to a new position; in other cases pieces of rock are torn directly from the channel.

Drainage basins

A leaf out of the book of plants

A **drainage basin** is the area of land that supplies water to a river. The number and arrangement of tributary rivers is related to the aim of moving rain as effectively as possible. You can understand how this is achieved by comparing the drainage basin of a river and its tributaries with a leaf

Figure 14.1 Drainage patterns
A leaf (a) has a network of veins which show many similarities with a river drainage basin, (b). Both systems try to perform their functions as efficiently as possible. You can often see **dendritic** drainage patterns eroded in soft bare soil (*right*); the deep channels are called **gullies**; exactly the same pattern would be formed on the slopes of a volcano, a mineral waste dump or any newly exposed land surface.

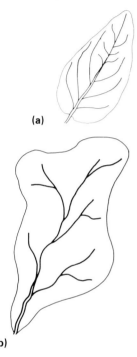

(Fig. 14.1). The plant has to supply nutrients to the leaf as effectively as possible; the river system has to remove water as effectively as possible. This pattern of water channels (called **dendritic** drainage) or leaf veins is very common in nature and is also used to supply blood to your body and oxygen to your lungs: it is a well tried and proven pattern. Sometimes, however, a river system flows over rocks with different resistance to erosion. It is quite common to find easily eroded rocks and more resistant rocks side by side in bands. Sedimentary rocks, such as shales and limestones, are often tilted and eroded to form **cuestas**, which are ridges with alternating steep (scarp) and gentle (dip) slopes. In these cases the drainage pattern becomes distorted and more trellis-like. The landscape becomes distinctive too (Fig. 14.2). The pattern of rivers is important because each river flows in a valley. Landscapes are made from patterns of valleys and so the whole character of a landscape can be affected by its rivers; and how quickly they remove rock fragments, even if it is an incidental by-product of water flow.

Exercise 14.1 Drainage basins

1. Select a page from your atlas that shows a mid-scale map of your home region (within 150 km of your home).
2. Choose a river that has at least several tributaries and choose one town beside the river.
3. Now trace the plan of the river and tributaries upstream of the chosen town. Is the pattern dendritic or is there some other pattern?
4. Draw a line around this network of channels to separate them from nearby river systems. The area inside this line is the drainage basin for the river upstream of the town.

Figure 14.2 Scarp and valley landscapes
This region consists of alternating bands of impermeable shale and permeable limestone. Many streams are required to carry the water away from the shale. Fewer streams are needed on the limestone, where much water sinks into the rock. The result is a **trellis** drainage pattern. Because the shale is eroded more easily than the limestone, the limestone stands out as ridges called **cuestas** and the softer shale layers make strips of lowland valleys.

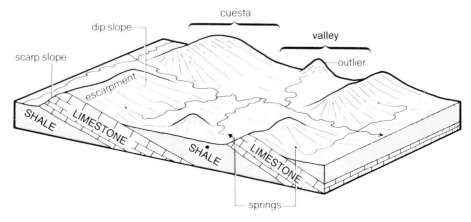

River flows

Too much of a good thing

We can learn much about river transport and erosion from newspaper reports. They tell us about extreme events when transport and erosion are at their most effective. The three examples below show this clearly. Each quotes an eyewitness account:

Colorado

❝ The water was running about normal or maybe a bit higher... I got curious about the creek..., it was raining you know. I got up to look out the front door and there it was coming up as fast as everything. By then it was already around the house. It started coming in under the door and I knew it was going to be bad, but by then it was too late to get out.

I put my dog on the bed and it started jumping around, floating here and there... I began to get awful cold and was shivering ❞ wet. It smelled awful, that mud and dirt in the water...

Nevada

❝ I started walking towards the coffee shop [and] I looked up for a second. I became disoriented because I thought the mountain had moved. Then I realized what we were seeing was a wall of water about 20-25 feet [7-8 m] high, stacked with cars, trailers, etc., smash into the coffee ❞ shop... and it exploded like there was dynamite inside...

South Wales

❝ It had been raining on and off for several days and the ground was thoroughly soaked so that it couldn't hold any more water. Then on Thursday it rained really hard all day. The sky was a dull grey and it was cold, so I stayed inside. I could just see the stream that flows through the village from my window. A pleasant little stream it usually is, bubbling along beside the road and disappearing under the bridge. But for the last couple of days it had become more and more swollen with rainwater and on Thursday it had turned a dull grey brown with mud. The water level was so high I thought it might wash away some of the brickwork from around the bridge. The water got higher and higher and pounded against the brickwork that protected the road. Then during the night I heard a loud crash. As soon as it was light I looked out of my window and saw that the bridge was gone and about 100 yards ❞ [91.4 m] of road washed away too...

Figure 14.3 Water power Waterwheels once powered the machinery for industry.

Besides experiencing floods, we often sense the hidden power of running water. We use running water to drive turbines and generate electricity; the first part of our Industrial Revolution was based on the power obtained from waterwheels. The wheel shown in Figure 14.3 is driven by water falling on slats and pushing each one down in turn. The water flowing over the wheel hardly looks up to the job of driving all the machinery of a textile mill or a forge. Nevertheless this is what such wheels did.

Using the evidence of waterwheels, perhaps we should revise our impression about energy available even in small streams, but we need little imagination to realize the real horror of the effects of floods on homes and people. The three eyewitness descriptions show that rivers can have many effects. The Colorado River is in a part of the southwest United States which is prone to intense storms. Flow over the soil surface is common and floods are expected. But even here storms can be unusually large as with the "cloudburst" that caused flooding in a canyon in Nevada. The Welsh story shows flooding due to just the opposite effect — a long period of steady winter rain that saturated the soils so that they could take no more water.

River transport

A hop, skip, and bump

Each of the stories on page 172 vividly demonstrates the power of a river in flood to break up and carry away most things in its path. In Wales a road was undermined and then carried away in pieces; in Nevada whole cars and trailers were tossed about like toys. But notice too the frequent references to finer debris in statements such as "It smelled awful — all that mud and dirt in the water." Clearly, a whole range of sizes of material was being carried along.

The descriptions tell us that the finer material was being carried along in **suspension** (Fig. 14.4), but what about the larger? To get some answer to that question we can turn to the mountains of Norway and an ice sheet that straddles the Arctic Circle. Here part of the ice sheet melts each spring and sends torrents of water down steep riverbeds to the nearby sea. You can stand over a kilometre away from some of these rivers and still hear the roar of the water as it cascades over bare rock steps in its path. And every few minutes the constant sound of the rushing water is broken by a deep thud, like a shotgun being fired. These are the sounds of huge boulders, once plucked from solid rock by glaciers but now being rolled over the edge of a rock step. Each thud signifies that several tonnes of boulder has just been pushed a little further by the flowing water.

Figure 14.4 Sediment transport
(a) Material in **solution** is invisible but may make up half the total sediment carried. (b) Clay particles are so small and light they can be kept in **suspension** as long as there is any water movement. (c) Sand particles are too big and heavy to be kept suspended in the water continuously; instead they *bounce* or *hop* along in the lower layers of the water when the flow is fast. (d) Particles larger than sand rarely leave the stream bed and are only **rolled** along when the water is moving very quickly. (e) Together, all of these materials make up the **sediment load** transported by the stream.

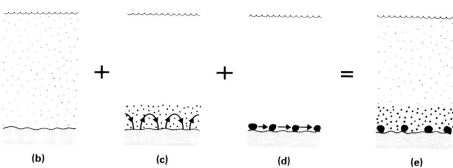

If water flows fast enough it can **roll** and even **bounce** boulders along weighing several tonnes. There was a famous case of this at Rapid City, South Dakota, when a flash flood, which followed a downpour over the Black Hills, littered a trailer park with boulders and shattered the mobile homes. But even a "normal" flood has enough energy to carry silt, clay, and sand into houses hundreds of metres away from the main channel.

Where does the transported material come from? The most obvious supply sources are the banks and bed of the channel carrying the water. Recent investigations of some streams draining through urban areas in smooth, concrete-lined channels give surprising evidence of this. Within a concrete channel streams have little chance of eroding new material. Channels can be eroded only when there is something rough for the water to push against. But while no new material is eroded, the rain from roofs and roads is continually added. As a result the stream gets bigger and acquires more energy, but there is nothing for the energy to be used on. When such streams re-emerge downstream of the towns, they are usually allowed to flow back into their natural channels. With such a lot of unused energy the streams then scour their beds and rapidly make their channels deeper.

Scouring is an important effect and it can be dangerous if it happens near bridges or buildings (Fig. 14.5). It shows how effective a river can be at eroding its bed.

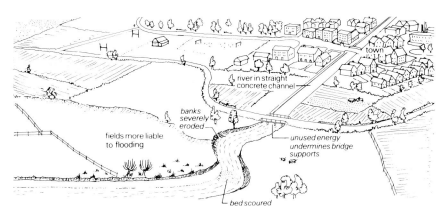

Figure 14.5 Scouring and undercutting of a river near a town

Because much more energy is available in time of flood, the larger material is moved at these times; sometimes with unfortunate effects if people have constructed buildings and roads in the way. But it is all too easy to assume that floods are the only important occasions for erosion. Measurements have shown that often just as much material is moved between floods by suspension and solution. This is because floods may happen on only a couple of days a year. For the rest of the time, the water may not carry so much material that it turns brown, but particles are still on the move!

River size

Bank robbery

Although extra energy causes scouring of the beds, it is worth remembering that banks are continually being eroded as well. Think back for a moment to the way hillside material is moved. Part is by solution, but the rest is by slow or occasionally fast movements of soil and rock down towards the river. If the stream did nothing about this, the channel would eventually be squashed narrower and narrower by these hillside movements. Often the river must be kept quite busy eroding its banks just to maintain a width suitable for its flow.

However, even if there is much energy to spare, the river does not become very wide. Measurements show that most river channels have a rectangular cross section: deep channels are always wide and belong with large rivers; narrow channels are less deep and belong to smaller rivers. There is always a balance among width, depth, and flow, although streams flowing over gravel are shallower than those over clay, because gravel cannot form steep-sided banks.

SUMMARY: WEATHERING, TRANSPORT, AND EROSION

Weathering occurs when a rock breaks down into smaller fragments, or decomposes but remains in place. When it is moved away, then erosion has occurred.

erosion = weathering + transport.

Rivers erode by breaking pieces of rock from banks or bed and then immediately carrying them away.

The difference between weathering and erosion can be seen in Figure 14.6. A large rock is being weathered into irregularly shaped sheets on its exposed upper surface. Gravity will eventually send the broken fragments sliding into the river, thereby causing the upper surface to be eroded. The underside is being constantly eroded and smoothed by abrasion from particles already in transport.

Rivers use their energy to move material in different ways depending on the size of debris. Small (clay) particles are carried along in **suspension**; larger (sand) particles tend to **bounce** or hop along the bed, provided the water speed is great enough; the largest (gravel and pebble) particles tend to be pushed and **rolled** along only at times of very fast flow. Of course, all the time and at every stage of the river material is also moved in **solution**.

Figure 14.6 Weathering and erosion of a granite boulder

Waterfalls

Mist, spray, and thunderous noise

In upland areas river channels usually have steep gradients and they often cascade over waterfalls or tumble over rapids.

In the Yosemite Valley (California), the Yosemite River sweeps round a bend and then plunges nearly 800 m over a huge granite lip before being dashed into seething foam on the rocks below (Fig. 14.7). This spectacular waterfall is impressive, but waterfalls are not rare: they occur throughout the world in upland areas. Waterfalls are found throughout western North America because of the ruggedness of the relatively young mountains of the Cordilleran region. Glaciers have increased the ruggedness by carving deep U-shaped troughs that have many smaller "hanging" tributaries from which streams plunge. On the western slopes of Mt. Robson, the highest peak in the Canadian Rockies, the first major tributary of the Fraser River flows out of the Valley of a Thousand Falls.

In the upper reaches of their courses, streams appear to have no control over their movements; they are thrown from one cliff to another, forced to create rapids or plunge through deep gorges. Yet in most lowlands there are neither rapids nor waterfalls. Such a contrast must have an explanation.

Here we are back with nature's fragile balance. A waterfall is an irregularity in the path of a river too great to be formed by the river itself. You can see this by looking at the nature of erosion at a waterfall. There is largely a "self destruct" mechanism at work. The lip is constantly eroding back in an attempt to give the river a more even gradient. How far streams succeed in destroying waterfalls depends largely on time and the nature of the rock.

But if waterfalls are not natural results of river action, how are they caused? One clue is that they are common in areas of faulting, where huge pieces of the earth's crust have been displaced vertically often hundreds

Figure 14.7 Waterfalls
Waterfalls, such as the Yosemite Falls shown here, are major breaks in the course of a river. They will eventually erode their valleys away.

Figure 14.8a Niagara Falls, New York, USA
These falls have retreated about 19 km from the edge of the Niagara escarpment which created the original falls after isostasy changed the drainage of the Great Lakes. The modern falls are at the head of a long, narrow canyon which might erode back to Lake Erie in about 25 000 years.

of metres above others. In these cases waterfalls can occur on large rivers: Niagara Falls, between Lakes Erie and Ontario, and Victoria Falls on the Zambesi River in Africa are two of the best known (Fig. 14.8 a & b). Many others are the result of ice action. We shall see later (p. 211) how ice can gouge out steep-sided valleys and give waterfalls to valleys such as the Yosemite.

Many smaller waterfalls show how rivers pick out a resistant band of rock. Horizontal bands of resistant rock are especially difficult for streams to erode. High Force (Fig. 1.16) is still trying to erode through a near-horizontal basalt sill. The different types of waterfall are summarized in Figure 14.9.

Rapids and potholes

Nature's slalom

Waterfalls are one feature of upland rivers, but if you walk along the narrow hiking trails of Capilano Canyon between West and North Vancouver, BC, you won't see waterfalls, but rapids with coloured poles hanging above the rushing river. Stop and look more closely and you may find kayaks shooting

Figure 14.8b Victoria Falls, Zambia-Zimbabwe, southern Africa
At this location, the Zambezi River plunges over a fault. Downstream the river follows a sharply twisting gorge cut deep into rocks of varying resistances to erosion.

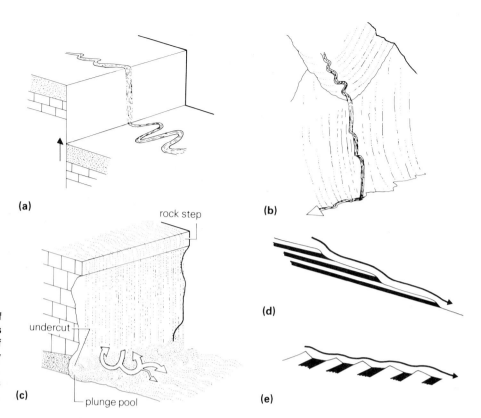

Figure 14.9 Types of waterfalls
Waterfalls can occur because of (a) a fault, (b) a hanging valley or (c) a resistant rock band. **Rapids** occur when rock dips (d) downslope or (e) upslope.

by, the paddlers doing their best to paddle a course between the poles while staying upright among swirling waters and rapids (Fig. 14.10).

The bed of the Capilano makes **rapids** as it flows over dipping rocks. The river has excavated some of the weaker beds and left the more resistant standing out as ridges. As a result, the water cascades over a series of irregular ledges. Most upland stream beds are irregular and usually you can see the rock they are made from. You will also see a variety of pebbles and boulders scattered about in places of slower-flowing water, waiting to be picked up during a storm and hurled against another part of the stream

Figure 14.10 Rapids
A kayaker shooting rapids.

177

(a) Potholes form when pebbles are caught in circulating eddies of a turbulent stream. The pebbles are whirled round by the water and they "drill" holes in the stream bed by abrasion.

Figure 14.11   Potholes

(b) Sometimes potholes become so large water has to spill between them as waterfalls.

bed. Sometimes such pebbles have produced distinctive circular forms called **potholes** as they have been caught in permanent water eddies and then used to drill vertical holes in the stream bed (Fig. 14.11 a & b).

We see features such as waterfalls, rapids, and potholes because the turbulent streams in an upland flush out most pieces of weathered rock and carry them away, exposing all the irregularities of bed and banks. So as the weaker rocks are eroded, the stronger rocks stand clear as the ledges of rapids and sometimes as the lips of waterfalls.

A deep incision   *Gorges*

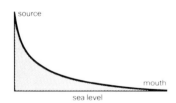

Figure 14.12   The line of a river channel from the source to the mouth
It is called a **long profile**. If it has no irregularities, as in this figure, it is called a **graded profile**.

If there were no hillside weathering and transport, a river would eventually cut a deep slot in the landscape (Fig. 14.13a). Sometimes when rivers cut down very rapidly we get close to seeing a slot in formation. It is called a **gorge**.

The Capilano River has carved a gorge (or canyon) into the granitic rocks at the southwestern edge of the Coast Range. The upper canyon has been dammed to create Capilano Lake, the main water reservoir for Vancouver. The lower canyon, over 100 m deep, is a park, and also home to a small, successful salmon hatchery.

(a) A gorge cut by a river where hillslope weathering is very slow. Notice the vehicle gives the scale.

(b) Capilano Canyon in Vancouver, BC, is a deep, V-shaped gorge created by an energetic river that plunges from the Coast Range to the Strait of Georgia.

Figure 14.13 Gorges

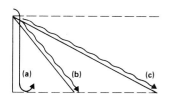

Figure 14.14 Spending stream energy
(a) A stream falling vertically over a waterfall uses most of its energy against the rocks at the bottom, causing undercutting and waterfall retreat.
(b) Falling steeply over a short distance, it uses a lot of energy and tends to erode its bed rapidly, forming a steep-sided valley. (c) Falling gently over a long distance, it has to spread its energy out and so can erode its bed only slowly. This gives gentler hillslopes a chance to form.

To cut gorges, like the Capilano, streams must lower their beds at a much faster rate than hillslopes have been able to form.

Streams can maintain steep courses only in headwater regions. Remember that the primary task of a river is to move water through the drainage basin as effectively as possible. So the river gradient must always be sufficient to take the water away from the headwaters and out to sea. In practice, this means that from source to mouth the river channel has a concave shape (Fig. 14.12). This is called a river's **long profile.** It gives steep gradients in headwater regions, but progressively more gentle gradients towards the river mouth. Of course, a gorge or steep-sided valley can be cut only when a river is using up its energy quickly, so these features are confined to upland areas.

Downstream the gradient is more gentle and a different set of valley features dominate the landscape, as we shall see. The point to remember throughout is that energy is used by the streams both in uplands and lowlands (Fig. 14.14). In uplands it is used to cut a channel quickly and leave steeply sloping hillsides; but in lowlands the energy is used in other ways. Some is used to transport sediment eroded from the uplands; some is used to erode more sediment from hillsides and the channel bed; and the rest is used by eroding sideways. The results are gently flowing and winding rivers in valleys with gentle slopes.

Topic Features of rivers with gentle gradients

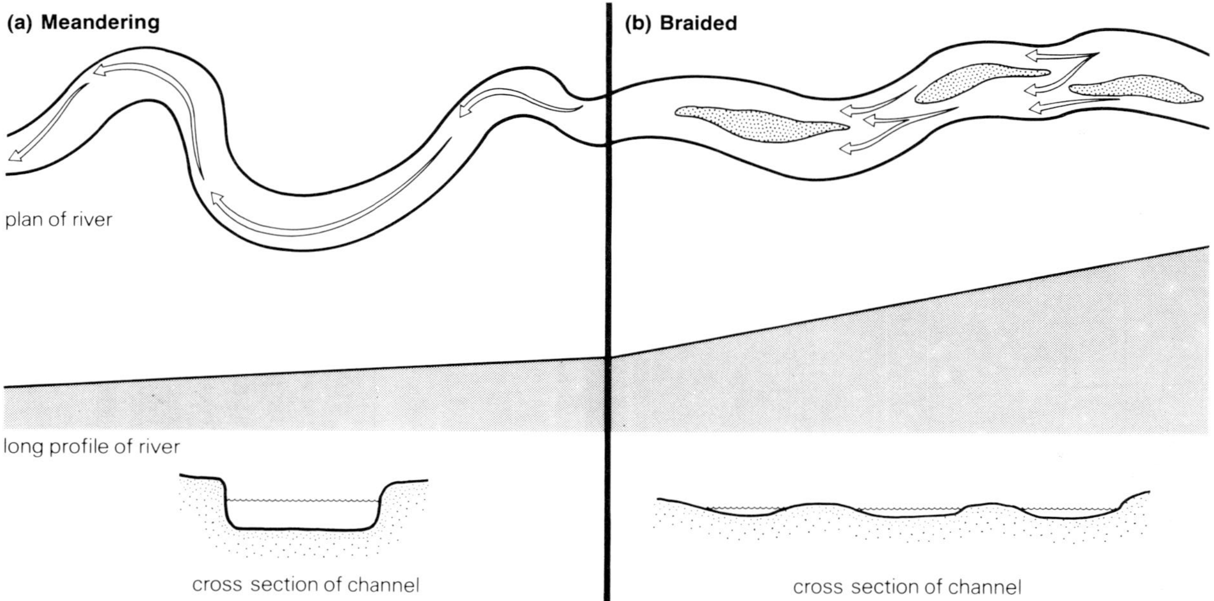

(a) Meandering

plan of river

long profile of river

cross section of channel

Figure 14.15 Meandering rivers have a deep single channel and flow over gentle gradients in clay alluvium.

(b) Braided

Braided rivers have a wide shallow channel split into many parts by small islands of sand or gravel. Flowing over slightly steeper gradients than most meandering streams, they are the downstream equivalent of rapids.

cross section of channel

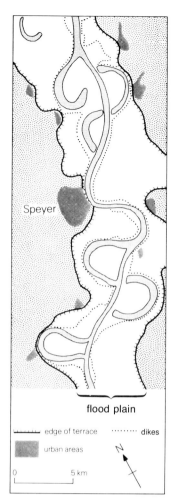

Figure 14.16 A section of the Rhine River in Germany Notice how the river has been straightened for navigation and how it has been diked to prevent water flooding onto the flood plain.

Swinging rivers *River meanders*

One of the most obvious features of a river flowing over a gentle gradient are its twists and turns. This is an essential part of a river system, as engineers have discovered. But there have been many attempts by engineers to ignore this rule. For example, the **meandering** (twisting) pattern of the Rhine seemed to be quite unnecessary to 19th century engineers who decided to build cuttings through many of the biggest meander loops (Fig. 14.16). When they had finished they had a much straighter course, but they discovered that the river now flowed even faster and this made hauling goods upstream even harder. Worse still, it scoured its bed and carried more sediment, but deposited it at the end of the "improved" reach. The engineers were left with a new problem: how to dredge the sediment away! Today the fastest part of the Rhine is bypassed by canal loops but the events that led up to its construction tell us that rivers flowing on gentle slopes do not meander in a haphazard fashion that can be altered at will.

If we stop to think about the speed at which water flows in a river, we shall see that it largely depends on the slope of the riverbed: steepen the slope (e.g. by cutting out meanders) and the speed will increase locally. The story of the river cuttings also demonstrates another point quite clearly. The Rhine has plenty of spare energy to cut down into its bed, but in its natural state it must maintain a balanced gradient to the sea. So the river cannot lose height too quickly. What then does it do with the spare energy? The answer to this seems to be that it meanders. Meanders make a river longer and also give the river something to do with its spare energy. Meandering rivers no longer simply erode their beds; they now have meander bends to erode (Fig. 14.15). This is in contrast to steep straighter rivers that use energy by **braiding** (repeatedly splitting).

Remember, too, that as with the rest of the river the meandering stretch is on a slope, so the whole meander pattern gradually "slips" downslope, eroding a flat strip of land called a **flood plain** as it does so (Fig. 14.17).

Step by step from source to sea *Flood plains*

Rivers with flood plains and meandering courses often flow in **alluvium**. This word describes any sediment temporarily deposited by a river and

Figure 14.17 Flood plains (1) Gravity pulls the whole meander belt downslope by making water erode the downslope sides of meanders. (2) With time a strip of land is eroded away. This is called a **flood plain**. (3) If the land slope is very gentle the meandering may become exaggerated and the meanders cut themselves off and reform.

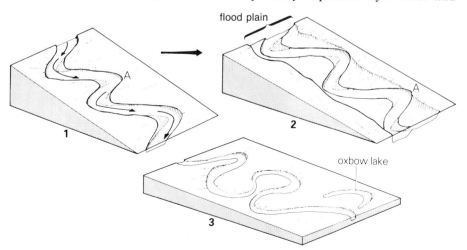

Exercise 14.2 River channel erosion

River erosion is complex because three things are happening at the same time: (a) the river is eroding its bed; (b) it is eroding its banks to preserve the channel width; and (c) it is following a winding path.

The winding (meandering) path is a feature of all naturally flowing water. Its effect is to cause uneven bank erosion, as shown in Figure 14.18.

1. What suggests that there is much erosion occurring on the outside bend of the meander?
2. What suggests that the deposition is occurring nearby on the inside bends?
3. Figure 14.18 also shows the line of fastest water flow. Trace the diagram and, using your answers to Questions 1 and 2, mark on places of erosion and deposition.

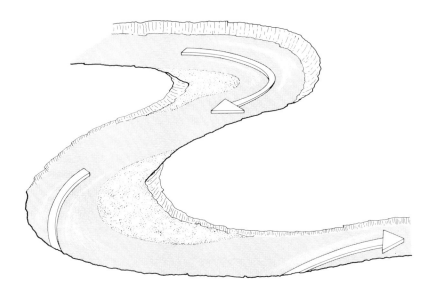

Figure 14.18 Features of a river meander
Notice that the energy in a meandering river is not evenly distributed, so it is quite possible for erosion to occur in one place and for deposition to occur nearby at the same time.

Topic Energy from rivers

We are increasingly harnessing the energy of even gently sloping rivers. Once this energy drove flour mills or small textile factories. Today it produces very large amounts of electricity for a multitude of users.

Figure 14.19 Our use of river energy

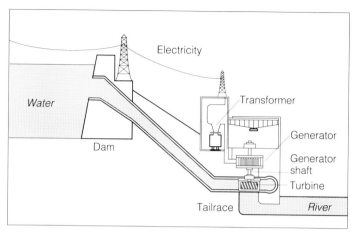

1 **Waterfall** e.g. Niagara Falls, Ontario

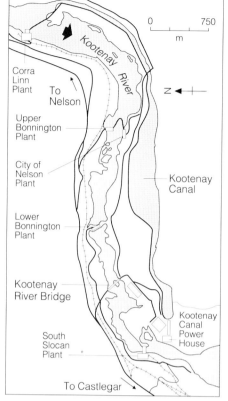

2 **Steep course of river** made into steps to create a head of water. e.g. Kootenay River, BC and 6 power plants. Similar developments: Tennessee Valley, USA; Rhine River, Switzerland and Germany; Snowy Mountains, Australia.

waiting in a flood plain or elsewhere to be transported to the sea. During a flood very little alluvium is carried right through a river system from where it was eroded to the sea. Most is deposited on the inside bends of meanders or flood plains. A particle of sediment may wait thousands of years before getting from the headwaters to the sea. In this respect the river acts rather like a long road with many sets of traffic lights. Great line-ups of rock particles develop at each set of lights (inside a meander bend) and have to wait while those at the head of the line-up move off. But these particles are mostly caught at the next set of lights. Although several thousand years is a fleeting moment in the life of a river valley, it provides time for the alluvium on flood plains to be reorganized by meandering rivers.

River terraces ## Steps of time

Rivers flow only because the land has been uplifted and a slope exists down to the sea. If land can be lifted up once, it can be lifted up many times. When land is lifted, rivers may flow over steeper gradients, have more energy even in lowlands, and so cut into their beds. But this does not stop them from meandering and eventually cutting new flood plains. However, a newer flood plain will be lower than the one eroded before uplift and it will be

Figure 14.20 The formation of river terraces
(1) Flood plain of the river before uplift. (2) Land is uplifted and the river cuts down into its flood plain. (3) A new flood plain is eroded away at the new, lower level, but ledges or terraces remain from the previous flood plain.

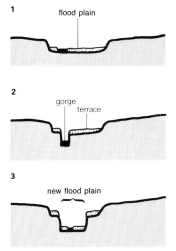

formed at the expense of the old. Eventually all of the original flood plain will be eroded away, but until it is, parts remain behind in the valley as "ledges" which are called **river terraces** (Fig. 14.20). River terraces provide dry flat land very close to rivers. As a result, they have been used extensively for the sites of towns and villages. For example, two opposing terrace remnants preserved near a river often make the task of river crossing easier. Similarly, a terrace beside the present river may provide the site for a port.

Exercise 14.3   People, meanders, and flood plains.

Figure 14.21 shows some human activities on a flood plain. Flood plains may provide good grazing land and flat areas on which to build houses, factories, and roads. They are also natural reservoirs in times of flood (p. 91). Natural flooding and people do not mix very well, and many attempts have been made to restrict the amount of flooding.

1. What is the purpose of the dikes?
2. Make a copy of the dikes and the river. Now try to complete the pattern of dikes that would be needed to protect the whole area. Remember that they are costly to build and that a short dike is therefore preferable unless it isolates a lot of land.
3. Why have people gone to the trouble of building dikes when there are no towns to protect?
4. In some places ditches have been dug across the flood plain. Explain why.
5. Suggest why flood plain ditches can have only a limited beneficial effect, even in times when there is no flooding.
6. Describe the route followed by the road. Explain why you think this route was chosen.
7. The bridge over the stream has many arches. Why might this sort of bridge add to the flooding problem upstream?
8. What special precautions would engineers have to take at the places labelled A, B, and C?

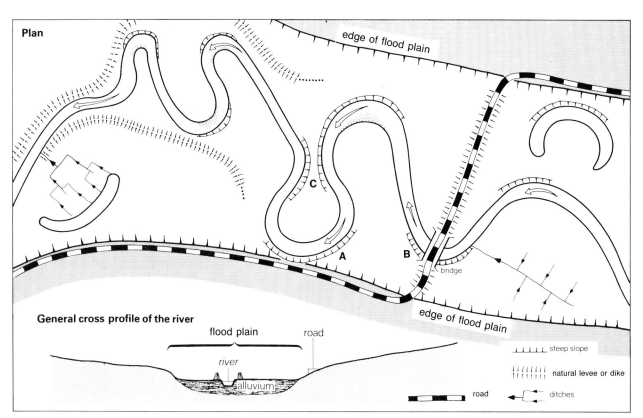

Figure 14.21 The plan of a meandering river

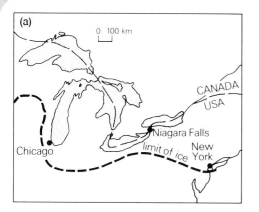
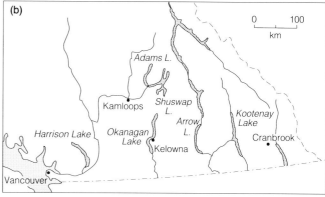
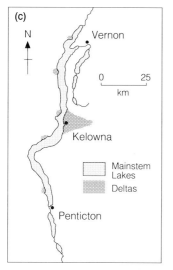

Figure 14.22 Lakes
(a) Filling a depression in the land surface caused by downwarping of the earth's crust. (b) Ribbon lakes filling valleys deepened rock basins in glaciated valleys in southern BC. (c) In the Okanagan Valley some of the mainstem lakes have been partially filled by deltas. (d) Okanagan Lake showing the city of Kelowna on its deltaic site.

Topic   Lakes

A lake is a natural reservoir of water formed in a depression in the earth's surface. Water in lakes is almost motionless. To transport sediment, rivers need to maintain a reasonable speed, therefore when they flow into lakes they usually deposit the material they are carrying. As a result rivers fill up lakes; they cannot create them. The biggest lakes are caused by the land sinking by folding (Fig. 3.7). The Great Lakes that form part of the boundary between the USA and Canada represent a downward fold of the earth's crust so big that the lakes are 240 000 km^2 (the size of Britain) (Fig. 14.22a). Similarly, **faulted landscapes** often create basins for lakes. Lake Baikal in the USSR, the Dead Sea and the Sea of Galilee in Israel and Lake Tanganyika in East Africa all lie within rift valleys.

Throughout the Cordilleran region of western North America **rock basins** have been created by valley glaciers, and later filled with water to form long, thin **ribbon lakes** (Fig. 14.22b). These lakes have maximum depths of 300 to 500 m, and very irregular bottom long profiles due to glacier action and the deposition of debris by the melting ice. Harrison Lake, near Chilliwack, BC, has a surface elevation of 10 m above sea level, but a maximum depth of 275 m.

Lakes are temporary parts of a drainage system. They will eventually be filled by sediment brought from upstream by rivers, but while they survive they are important in controlling floods. Lakes store up the flood of a river with perhaps only a few centimetres' rise in level. This surplus water escapes at the downstream end slowly and in a more even flow. We have learned from this, and many of our flood control schemes now include artificial lakes **(reservoirs,** p. 98).

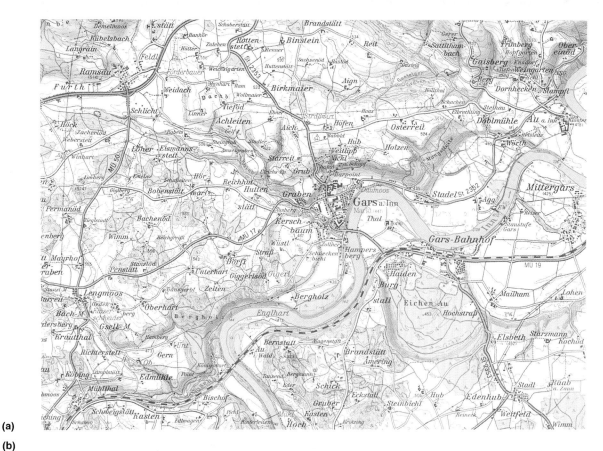

(a)

(b)

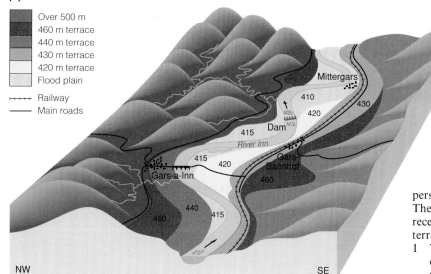

Figure 14.23 River terraces and the flood plain of the River Inn, South Germany

Exercise 14.4
River terraces and people

Figure 14.23a is a topographic map of part of the valley of the River Inn in southern Germany. Figure 14.23b is a perspective sketch of part of this area. The Inn is a meandering river which has recently lowered its flood plain. Many terrace fragments have been left behind.

1. Trace, from the map, the river's course and colour green the flood plain, about 410 m above sea level.
2. Mark on your tracing the small dam between Gars-Bahnoff and Mittergars. What height is this dam? How might the dam affect the gradient of the River Inn and the flood plain?
3. Show the three towns of Gars-a-Inn, Gars-Bahnhoff, and Mittergars. Briefly describe their sites.
4. Add the railway and main roads to your map. Compare these transportation routes.

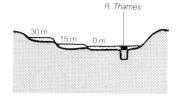

Figure 14.24 The River Thames in London flows on 20 m of alluvium

Rising rivers

Uplift is not always the rule. Sometimes the land sinks or the sea rises. This happened quite recently in geological terms (18 000 years ago) when ice from worldwide ice sheets (p. 204) melted and released vast quantities of water back into the oceans. With a higher sea, rivers had to adjust to a new, more gentle gradient. They did this by depositing alluvium in the lower parts of their valleys. Now, for example, the River Thames in London no longer flows on top of solid rock but on a thickness of 20 m of alluvium (Fig. 14.24). In similar fashion, the Tigris and Euphrates rivers have extended their deltas well into the Persian Gulf, and the Colorado River has extended its delta into the Gulf of California. Most rivers are still depositing sediment in their lower valleys.

Many events have altered the shape of a river valley. Terraces show how the land has risen quite quickly many times. However, the fact that many rivers are now flowing on alluvium and not eroding their beds at all shows that sometimes the erosion of a landscape can temporarily go into reverse!

Exercise 14.5 Levées

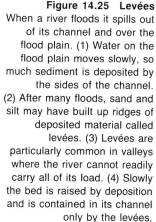

Figure 14.25 Levées
When a river floods it spills out of its channel and over the flood plain. (1) Water on the flood plain moves slowly, so much sediment is deposited by the sides of the channel. (2) After many floods, sand and silt may have built up ridges of deposited material called levées. (3) Levées are particularly common in valleys where the river cannot readily carry all of its load. (4) Slowly the bed is raised by deposition and is contained in its channel only by the levées.

Rivers build up **levées** (natural dikes either side of the channel) by depositing sediment brought from upstream (Fig. 14.25). Sea level rises and the river has to raise the level of its bed and banks in step. It is a dangerous condition for people living on flood plains because the river flows *above* the plain, held in only by its levées.

The Mississippi River in the USA is noted for its large levées. The city of New Orleans has grown up beside the river, where no town ought to be. It happened to be the first almost-dry spot found by the early French explorers at the end of the 17th century. Only meant as a temporary settlement, today the city curls around a huge meander of the river, its sole protection being levées, most of which are about 8 m high. The people of New Orleans often say they used to look up at the water, before pumps were installed to push out the rainwater, and jested that people there were born with webbed feet! It is a very good example of the struggle between people and their environment.

1. If the levées are built of river clay and silt, are they strong and stable or liable to breaching?
2. What are the dangers of building a town next to a river flowing within levées?
3. What precautions should people take in order to avoid the dangers you have identified in Question 2?
4. New Orleans is built on the outside bank of a meander where there is deep water for a river port. What causes deep water in this position (p. 181)?
5. Explain why a position on an outside bank might be most dangerous when a river is held in only by levées.

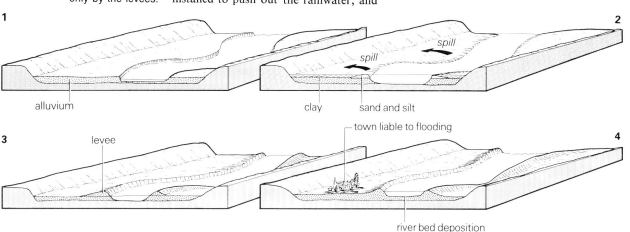

Topic Valley shapes

(1) Rivers normally just erode their beds. Sometimes this can be seen very clearly when the processes that make hillsides do not work very quickly. Some desert areas show these conditions: plateaux cut into by deep gorges. One of the best examples is the Grand Canyon (Arizona, USA) (Fig. 2.1)

(2) In many humid upland areas precipitation falling on the surrounding drainage basin causes rapid physical and chemical weathering. Loose rock and soil move down to the stream and some of the energy of the stream has to be used to remove it. Nevertheless many upland streams still erode their beds quickly and valleys have steep, straight sides.

(3) With less swiftly flowing streams, weathering produces more debris from the hillslopes beside it, and much more stream energy is used in transport. Channel downcutting into the debris is now slower. Soil is removed by solution, landslides, and mudflows rather than rock avalanches.

(4) When streams are near sea level, vertical erosion slows down and flood plains form by lateral erosion. Hillsides continue to deliver soil to the streams until they too have been reduced to gentle slopes.

(5) Rivers in lowland areas cannot erode vertically, and hillslopes deliver soil to streams slowly. Landscapes change extremely slowly in these conditions. Typical lowland areas include the Fraser Valley, the Mississippi and Amazon basins, the Netherlands, and the Nile delta.

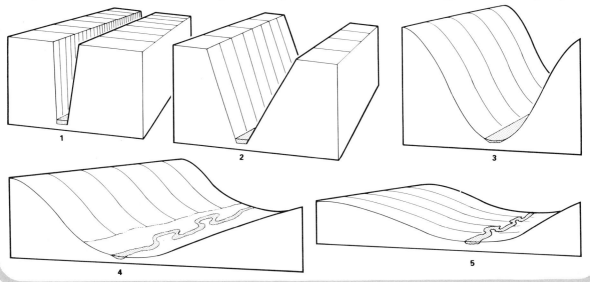

Figure 14.26 Valley shapes

Deposition in valleys

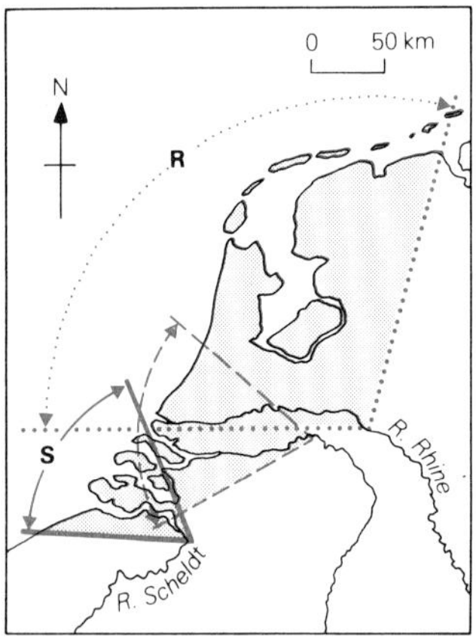

Figure 14.27 The Rhine (R)-Meuse-Scheldt (S) deltas

Alpha, beta, gamma...

When the Dutch set about building the dam that closed off the famous Zuider Zee earlier this century, they knew they had to find millions of tonnes of sand. It was no use looking to the land for supplies — they had little enough land already without dumping some in the sea! The only answer was to dredge the sand from offshore. In doing this they scooped back some of the sediment that had been stripped off in the past. In fact much of the Netherlands and part of Belgium are really above sea level only by courtesy of the Rhine, Meuse, and Scheldt rivers. Most of the Netherlands is part of a huge fan-shaped pile of sediment formed as the rocks of the interior of Europe were carried by these rivers and dumped into the North Sea (Fig. 14.27). The regions of sediment accumulation are called **deltas.** Each year millions of tonnes of sand, silt, and clay are pushed into seas and lakes by rivers, usually to build up deltas.

Some deltas are triangular; the Greeks even used the letter (Δ) to describe the triangular shape of the Nile delta in plan; but many can be found that are all sorts of odd shapes. This is because deltas are made from sand, silt,

Delta formation

Down in the dumps

When a river enters a sea or a lake it sends a current of water from the shore containing all of the sediment eroded from its drainage basin (Fig. 14.28). The current may extend for several kilometres away from the coast —

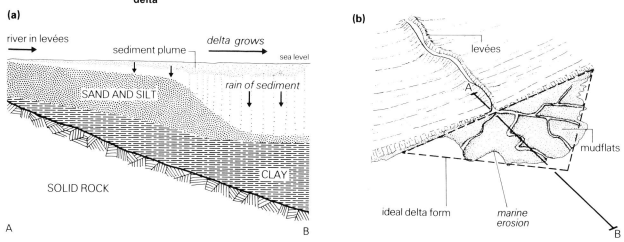

Figure 14.28 Growth of a delta

fresh water from the Mississippi can be detected 50 km out to sea — but all the while the water is slowing down so that the coarser sediment quickly drops out and "rains down" onto the seabed. The first material to be dropped is sand, then silt, with clay particles (being so small and light) settling farthest from the shore. Most of a delta is made from silt and clay; in the Mississippi delta only 2% is sand, and even in rivers flowing quickly from mountains, coarse sediment is less than half the total (e.g. the River Po in Italy flows from the Alps but the sand content in the delta is less than 40%).

Small rivers push out a narrow current of water and create deltas that stand out from the coast. Because of this they are rapidly eroded. Large rivers push out a delta which is very broad and has a very gentle slope. Waves cannot erode this as effectively and so it remains as a prominent coastal feature.

A delta grows into a fan shape by depositing material out in the sea and then flowing over this in a channel formed by levées. From time to time these levées are breached (especially on meander bends during floods) and the river finds a new route to the sea. There have, for example, been six major new delta growths in the Mississippi in the past 5000 years alone. In between these channels, mud flats form which are rapidly colonized by either grass (salt marshes) or in tropical regions by mangrove trees (mangrove swamp). When only part of the delta is growing, the rest is eroded by the sea and this gives the delta its irregular shape, as though something had been gnawing at it.

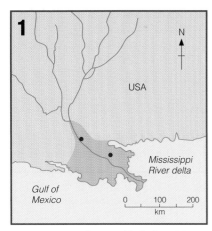

The Mississippi forms a bird's-foot delta on the Gulf of Mexico's isostatically sinking coastline.

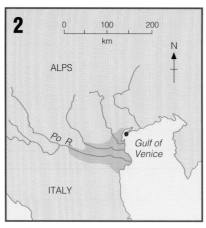

The Po in northern Italy has created the Plains of Lombardy and a large delta with many islands.

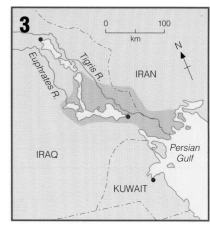

The Tigris and Euphrates rivers have a vast delta, formed within an estuary, on the Persian Gulf.

Figure 14.29 Deltas of the World

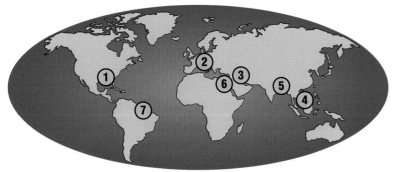

Exercise 14.6 Deltas of the World

1. Through an atlas search, identify a major city and major agricultural activity in each of these deltas.

2. Draw maps of the deltas of the Fraser and Mackenzie rivers. Compare the formations, areas, and land uses.

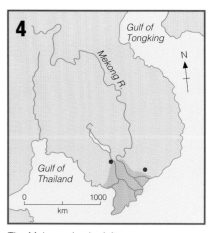

The Mekong river's delta covers over 10 000 km² of southern Vietnam.

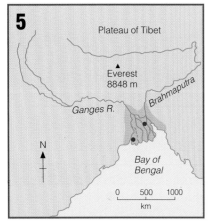

Bangladesh occupies most of the fertile, but often flooded 50 000 km² delta of the Ganges and Brahmaputra rivers.

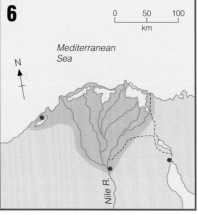

The Nile's bow-shaped delta extends over an area of about 4500 km².

The delta of the Amazon, Xingu, and Tocantins rivers is filling a great estuary.

15 Running water and limestones

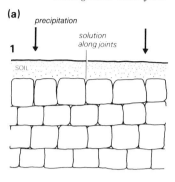

Underground water flow

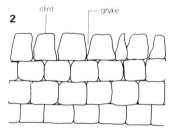

Figure 15.1 Limestone surface features
How a limestone pavement forms: (1) The limestone block surfaces are attacked by acidified rain water. The limestone rock surface is now very irregular but this is hidden by the soil cover. (2) Glacial erosion strips off the soil cover and exposes the limestone pavement.

The inside story

If you visit Castleguard Cave in the Canadian Rockies, the Carlsbad Caverns in New Mexico, USA, or the Yorkshire Dales in northwest England (Figs. 15.2, 15.4, 15.5, & 15.6), chances are that you will see small groups of people clad in muddy rubber suits and carrying miner's lamps, helmets, ropes, and wire ladders. These people explore caves. Through their investigations we have learned much about the way **limestone** scenery develops.

Limestone is a word used for a whole range of materials from the soft white chalk of southern England to the massive blocks of "mountain" limestone found in the Rockies, and Kwangsi province, China. But it is to the "mountain" limestone that we first turn our attention, because it contains the caves and passages that have made limestones so well known.

Like all limestones, the "mountain" limestone is mostly made from calcium carbonate, with very little other sediment. Calcium carbonate is particularly vulnerable to reaction with acidified rain water and stream water. Indeed it is so reactive that most limestone solution occurs on the surface before the water even gets underground.

Chemical weathering attacks the surfaces of materials and its effectiveness depends on the number of joints in the rock. Mountain limestone may have joints only every metre or so. In a way this concentrates the weathering and causes the few joints to be attacked vigorously (Fig. 15.1). The results of this weathering are sometimes exposed as a **limestone pavement.** Although each joint may start out being just a hair's breadth, it soon widens sufficiently to allow water to run through it quickly (Fig. 15.2).

Figure 15.2 Erosion in limestone
Water rushes through a bedding plain in limestone. Notice the pebbles below the waterfall. These are now the main tools of erosion.

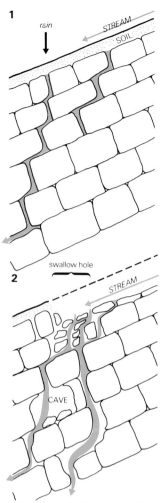

Figure 15.3 How a swallow hole forms
(1) Precipitation passing through soil becomes acidified. Water percolates down joints, dissolving the faces of the blocks. A stream runs over the area, providing more water which will sink right through enlarged joints. (2) Blocks near the surface become dissolved away and they collapse, providing a depression (swallow hole) down which the stream sinks. As soon as the joints between the blocks become wide enough, running stream water carrying sediment causes rapid erosion, producing passageways and caves. Notice how the water can move independently by several routes. This is what makes cave systems into three-dimensional mazes.

Once running water enters a limestone, joints are widened much more rapidly and the effects of solution become much less important. Running water, together with the sediment load it carries, rapidly abrades the joints and makes tunnels — natural drainpipes — through the rock. Occasionally tunnel erosion causes roof falls and a **cave** is born (Fig. 15.3). Solution continues to be important near the surface, dissolving blocks in such a way that they collapse and give depressions called **swallow holes** (Fig. 15.4). But underground there is very little solution. Indeed much dissolved limestone is precipitated as the **stalactites** and **stalagmites** that make caves so beautiful (Fig. 15.5).

Stalactites and stalagmites are the cone-shaped pinnacles of white calcium carbonate shown in Figure 15.5. Stalactites are formed as drops of water seep through the joints in the upper surface of a passage or cave. The water coming out of the joint is saturated with calcium bicarbonate, but when it comes out into the cave or passage much of the carbon dioxide gas goes into the air. This leaves the water with a reduced ability to keep calcium carbonate dissolved, so some precipitates out as calcium carbonate. Precipitation is very slow and stalactites may take thousands of years to form. Note that perfect cones are unusual. If, for example, the surface on which the drip forms is sloping, the water will run along the rock depositing calcium carbonate in a line. Eventually this builds up a flat sheet of stalactite called a **curtain.**

Water dripping from the roof of a passage or cave will fall onto the floor where further precipitation may occur. In this case, a rough, irregular cone of calcium carbonate grows upwards from the floor to produce a stalagmite.

Sometimes water runs over the sides of a cave and then precipitation coats the rock in flowing white **flowstone.** Streams also deposit flowstone, or sometimes create little dams which retain water. The range of forms is endless!

You can form an idea of the nature of an underground limestone landscape by a detailed look at a particular cave system in southwestern England. Swildon's Hole comprises over 8 km of accessible stream passage below the Mendip Hills. The stream sinks into a swallow hole in a field and comes out again at the famous show cave at Wookey Hole, near Wells. The straight-line distance between swallow hole and spring is 3 km and the vertical difference 200 m. However, even the accessible passageway length is 8 km and there must be three or four times as much that people can't squeeze into, so this gives some idea of the extent of the three-dimensional maze that underground streams have eroded in the rock.

The description below is intended to give an impression of the nature of underground drainage systems from the viewpoint of the caver. Swildon's Hole cave system and other cave systems can be dangerous to the inexperienced. They are liable to flood in wet weather and there are deep drops.

On no account should caves be entered without proper equipment and an experienced guide. If you want to see a cave, you should visit one of the many safe show caves that are open to the public. Any other form of caving should only be done through a recognized caving club.

"Swildon's Hole is entered feet first between a pile of large boulders (the swallow hole) in the bed of a small stream. An intricate network of meandering stream passages lie ahead. The stream first falls over a 3 m drop in a passage 60 cm wide and 3 m high. This stream tumbles steeply

Figure 15.4 The entrance to Gaping Ghyll cave, Yorkshire, England

Figure 15.5 Some of the features of a well decorated cave in limestone
Below. Flowstone and curtain stalactite, the most common type of precipitate.
Below right. The less common cylindrical stalactites and stalagmites. *Right.* Cross section through a cave.

down towards the first cave, but the passage height quickly lowers so that a couple of crawls through water are needed. Eventually the streamway opens out into a cave 10 m wide, 20 m long, and 8 m high, with some stalactite formations but many fallen boulders. The cave continues through a passage some 2 m wide in places and 8 m high, which leads on after 100 m to a wet drop and then an awkward 4 m climb down the side of a pothole. Following the stream past much flowstone decoration soon leads to the head of a 7 m pothole which needs wire ladders and safety lines for descent. After climbing down the ladder and through a waterfall it is noticeable that the pothole has rounded boulders in the bottom. A little further on and there are two more potholes — a fine pair of water-worn circular features about 5 m in diameter...''

This description is a small part of the cave system that contains a stream, but there are kilometres of passageways that are now dry and abandoned by the stream. These are parts of the underground networks that were formed at an earlier stage when the outlet for the stream was higher.

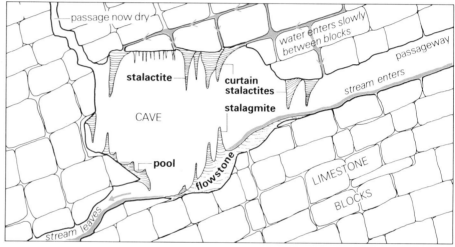

Limestone landscapes compared

Spot the difference

You might think that there ought to be much similarity between chalk and "mountain" limestone landscapes. The rocks are both made from calcium carbonate, are highly reactive with rainwater, and have no surface drainage; all the water flows underground leaving abandoned dry valleys or gorges. Yet the landscapes are very different — chalk landscapes have smoothly rounded hills and dry valleys with gentle slopes; "mountain" limestones have sharp angular outlines with vertical-sided gorges. The difference is in the nature of the rock. "Mountain" limestone is a physically hard rock that will not let water pass through; its permeability is due to the interconnection of cracks. which divide the rock into large blocks. In fact, the cracks are so regular that you can think of limestone as a pile of bricks. There is plenty of support for stable passageways and caves underground. By contrast chalk is physically soft permeable rock, without well-defined joints; instead it is crossed through and through by multitudes of hairline cracks. Chalk cliffs along the coast are steep only because they are being eroded rapidly by the sea.

Figure 15.6 Ingleborough Hill dominates the complex limestone landscapes of the Wensleydale area of northwest England.

Exercise 15.1   Wensleydale

1. Winterscales Beck disappears at 744784. Give the grid reference where it reappears as the River Doe.
2. How far does the water travel underground between the points noted in question 1?
3. Study the 56 potholes that surround Ingleborough Hill. What do they and the streams noted above tell us about the main stratum of limestone?
4. Draw and label a cross section from Gaping Gill (751727) through the summit of Ingleborough Hill, to the top of Combe Scar (730797).

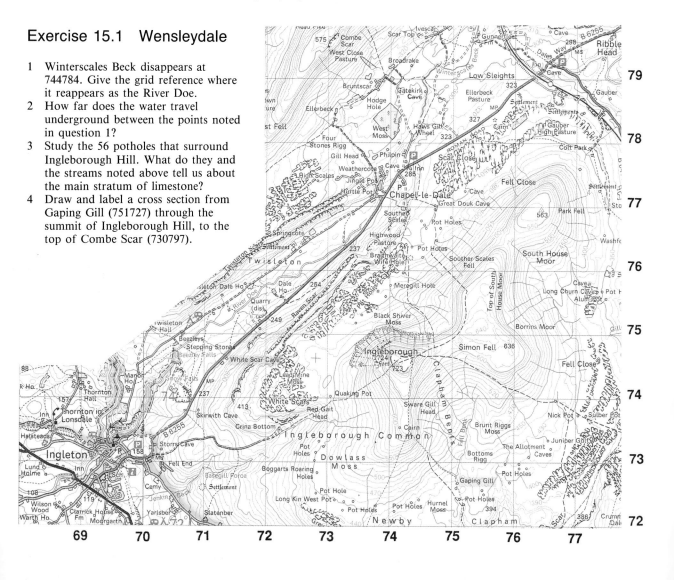

16 Deserts

Flash floods

The discovery of Lucy

One night Lucy was sleeping on a dried-up river bed in semi-arid Ethiopia when a rainstorm broke over the nearby mountains. Lucy neither heard nor saw the storm, but as the rain gathered into torrents and flooded out of the mountains she was already doomed. Soon this sudden flood of water (**flash flood**) was racing down the dry river bed to engulf and bury Lucy and her companions in a mud and water mixture like warm, wet concrete. That was three million years ago. Lucy is the oldest known "human" skeleton found on earth. Named after a Beatles' song, she may have been one of the first to die in a sudden desert flood (Fig. 16.1), but many people have suffered the same fate in recent times.

Deserts (arid lands) are areas where so little precipitation falls that vegetation survives only in specialized forms and it covers a very small fraction of the land surface. Along the desert margins are the semi-arid lands where precipitation is greater but still not sufficiently plentiful to support an unbroken vegetation cover. There is no easy way to define an arid or semi-arid area. It is best attempted in terms of vegetation, although some people have defined a desert area as having less than 250 mm of precipitation a year.

Arid and semi-arid lands cover 30% of the earth's surface and are inhabited by 600 million people, so although these lands may be unfamiliar to us, they are important and should be understood.

Alluvial fans

Fans of the mountain front

About 8 km from the town of Suez in Egypt a gently sloping gravel plateau (**hamada**) ends abruptly at the rising front of a mountain range (Fig. 16.2). This sharp contrast in landscapes is typical of deserts. The mountains are

Figure 16.1 A rare photograph of a flash flood moving down a desert valley In the foreground the stream bed is totally dry and there are no storm clouds in the sky. The flood has been caused by a local storm somewhere in the mountains. In some parts of the USA, flash floods are called **gully washers**.

Planned development

░░░░ houses
░░░░
░░░░

▦ industry

▩ shops

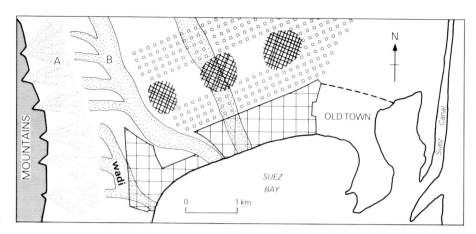

Figure 16.2 The Suez area, Egypt

there because sometime in the past a huge faulted block has been tilted and thrust up 800 m above the surrounding plains. Now these mountains tower over the Gulf of Suez, their stark edges bitten into by weathering and erosion caused by the run-off from occasional torrential rain over thousands of years.

Storms are rare, perhaps occurring only a couple of times a year, but each storm causes water to cascade over the steep slopes and carry away broken rock debris that litters the slopes. On the floors of the valleys (called **wadis**; Fig. 16.1) water and debris combine to cause torrents that surge and boil with clay, sand, and pebbles. Streamflow in a wadi is very like warm, runny, liquid concrete with lumps in it. Yet the rain is so short lived and the ground so parched that the water barely has time to pick up the debris, let alone move it very far. On steep hill-slopes water has enough energy to carry very coarse debris, but valley floors are much less steep and the plains beyond the mountains even more gentle. As a result much of the coarser material is dumped at the foot of the mountains as huge **fans,** leaving only the finer debris to be carried on across the plain. These fans are enormous: many are more than 300 m high and stretch out several kilometres

Figure 16.3 Alluvial fans stretching out from a mountain front
There is a road in the bottom left-hand corner of this photograph. The distance from road to top of fan is about 3 km.

from the mountain front (Fig. 16.3). Very often they spread into one another to form a continuous "fringe" of debris (**bajada**) around a mountain range.

Storms in the mountains are dangerous because they arrive with little warning. People have been drowned when venturing into a dry and apparently safe wadi, only to see a wall of water roaring down upon them! Roads are destroyed by flash floods, and damage can be very costly. Both Israel and Jordan have suffered from roads being washed away in the past few years.

Pebbles and plants

Wadi floors

At first light every morning women from the rural villages near Erfoud in Morocco make their way out across the nearby barren and pebble-strewn plateau. Slowly they guide their herds of goats to places where a few hard and dry plant shoots might still survive (Fig. 16.4). All day long they wander with the goats under the blazing sun, but they are careful to keep to the barely discernible broad and shallow channels in the pebbly desert.

The importance of these channels becomes very clear only a few times a year. At these times heavy convectional storms sweep over the area and deluge the plateau and nearby mountains. For a few hours water cascades from the mountains and floods out over the plateau, filling the shallow channels on its way to the sea. Almost immediately after the rain starts, the tiny hard and shrivelled buds that have been hiding between the pebbles swell and burst into life. Now the landscape takes on a green haze as millions of desert plants come into leaf and flower.

Desert flowers are typical of life in arid regions. For months, perhaps years, there is no water, no erosion, no plant growth. Then an occasional violent storm causes the whole landscape to burst into life — rivers flow, rock debris is washed from slopes and swept along river channels, and plants spring up and bloom. Suddenly the storm waters have gone, absorbed by the pebbly alluvium, evaporated by the blazing sun, and within a few hours the landscape is once more parched and dry; the flowers begin to fade only a few days after the rain.

After a few days the brief period of flowering is over, the seed scattered, and the desert apparently lifeless once more. But because more water sinks into the channels than elsewhere, this is where most plants survive most easily, even though survival for the majority is as seeds. Consequently this is where the herds of goats wander in the search for the scant browse on which their lives depend.

Figure 16.4 The gravel-strewn plateau over which flash floods sometimes flow The goatherd is to be seen in the middle distance collecting water from a well.

Figure 16.5 The shallow channel that signifies the track of a flash flood
Notice the road is built straight across this route, which gives some indication that flooding is very rare.

Exercise 16.1 Water erosion in the desert

1. Figure 16.1 shows a wadi with a rare but well known desert feature. It is a flash flood. Explain why flash floods occur. How can you tell that this is a photograph of one?
2. When a flash flood flows down a wadi it moves over sand and pebbles such as are shown in Figure 16.5. What do you think happens? What might happen to the car if it were not moved clear?
3. Why do alluvial fans occur at the edge of the mountains?
4. Flash floods travel only a short distance over plains because the supply of rainwater soon ends. What happens to the floodwater and the sediment it is carrying?

Topic Inland drainage basins

Because storms are so rare and restricted in area, much stream water never reaches the sea. In many areas, especially those where large-scale faulting or deflation have produced enclosed basins, storm water builds up into temporary lakes, called **playa lakes** (Fig. 16.6). These lakes quickly evaporate, leaving behind dissolved salts. Thousands of years of evaporation has formed "salt lakes" which may have salt hundreds of metres thick accumulated in them. The Great Salt Lake, and Bonneville salt flats, Utah (USA) are perhaps the most famous of these.

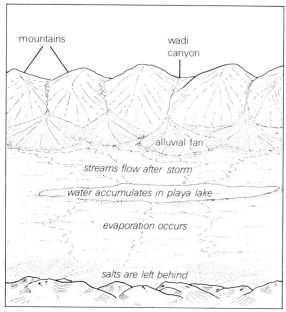

Figure 16.6 Salt Lakes
Playa lakes are salt lakes formed when local drainage basins dry up.

Exercise 16.2 Urban development in the desert

At the southern end of the Suez Canal lies the town of Suez. It is on a flat, almost featureless strip of ground about 5 km wide, covered by sand and pebbles (Fig. 16.2).

The original site dates back to ancient history but the plan to establish a new city will cause considerable development. Houses and industry will be built on the previously unoccupied sand and pebbles. Civil engineers need to know whether the new city will be built over any problem areas.

1. Look at the map (Fig. 16.2) that was prepared for part of the area. At A there are some spreading channel features. What do you think they are? The material in which the channels are cut is sand and pebbles, so of what do you think the area at A forms a part?
2. Water leaving an alluvial fan is very unpredictable, but there are some areas where it is more likely to flow than others. They are mostly marked by broad shallow channels such as that shown at B. What are these channels called?
3. Compare the plan for the expansion of the town with the position of the most likely route for water from flash floods. Describe what you notice. Can you suggest modifications to the plan that would help prevent houses and factories being built in the path of a flash flood?

SUMMARY: FEATURES PRODUCED BY RUNNING WATER IN DESERTS

The main features of an arid area are that (a) weathering is due to exfoliation (see Ch. 9); (b) precipitation occurs in occasional, intense but short-lived storms; and (c) there is no soil to store water. Water cascades over the bare valley sides and reaches channels quickly as **flash floods**.

Without constant moisture and a vegetation cover, rock fragments cannot easily be weathered down to clay size. Weathered debris littering slopes is therefore coarse — mostly sand, pebble and boulder size. Valleys in desert or semi-arid areas are called **wadis** in Africa and **canyons** in North America. Where wadis meet lowland areas the river gradient becomes less steep and the rivers are no longer confined to narrow channels. As a result, the river water spreads out and is able to carry less debris. At junctions of upland and lowland it is usual, therefore, to find large **alluvial fans**. If the wadis spill water into enclosed basins, **playa lakes** form (Fig. 16.6).

Sand dunes

Waves that never break

Rocky mountains and pebble-covered plains provide one type of desert landscape. But there is another type of desert landscape where wind-blown sand collects. These are the **sand seas** which are called ergs in northern Africa.

Sand dunes are probably the most famous of desert features even though they occupy less than a quarter of desert lands. They are really only piles of wind-blown rock fragments, yet to many they are beautiful and intricate patterns of smooth flowing shapes, like frozen waves in a golden sea (Fig. 16.7). In some places, such as Libya, sand seas cover vast areas and

Figure 16.7 Sand sea, Sahara Desert, Algeria

provide the world's most impenetrable and hostile land. These are the lands of the Sahara — the Grand Erg Occidental and the Grand Erg Oriental with their camel trains, Bedouin tents, and abandoned forts of the French Foreign Legion. But elsewhere they tend to occupy very small areas.

Sand transport

Borne on the wind

People with much experience of desert life can sometimes tell when a dust storm or sand storm is coming, but for most people they arrive without warning. Often the first sign is a change in the direction of the breeze. Next, dust will start to lift off the ground in an almost ghostly way, just as though someone had disturbed the floor of a haunted house. After a slight but noticeable rise in the wind speed, sand begins to move. At first sand blows in wispy trails over feet and around ankles. Then a brilliantly clear and hot day suddenly becomes obscured by a dark cloud stretching from the ground high up into the sky and getting nearer every second. The darkened sky is not a sign of precipitation, but of parched hot air, dust, and sand. The dust (silt and clay particles) gets into the lungs, hair, eyes, and ears; the sand beats against any exposed flesh with a force that can be painful. The wind has become stronger now, causing violent swirling around any obstruction. It is now taking sand and dust into every nook and cranny and piling it up against rock, wall, and fence alike.

A mass of air blowing across a landscape behaves in a very similar way to a river. In the desert, air moves quickly over an irregular bed of sand and pebbles. Any loose material is picked up and moved in suspension or by bouncing or rolling. The faster the wind blows, the more energy it has to move bigger pieces of material and the greater the total amount of debris transported (Fig. 16.8).

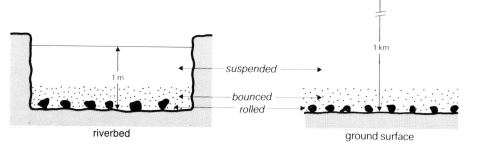

Figure 16.8 The movement of material
Material is moved by wind and water in exactly the same way. Only the height to which suspension and bouncing can occur varies.

The similarities between wind and water do not end with transport. Water carries away weathered debris or abrades its channel by continually rolling and bouncing stones across it; wind abrades by driving sand across any exposed rock. There can be no doubt about the abrasive efficiency of sand-filled air — the technique is often used to clean buildings. Nevertheless, a jet of air and sand from a compressed air line is vastly more powerful than the wind and we must understand that, in natural circumstances, sand abrasion works very slowly. By contrast, dust does not abrade and so natural sandblasting is confined to the height that sand can be lifted by the wind — about a couple of metres. Of course, the larger pebbles are rarely moved, so that eventually differential removal of sand, silt, and clay particles from the surface of a desert may lead to the development of a stone pavement or deflation armour (Fig. 16.9).

Figure 16.9 Stone pavements
Stone pavements may develop when wind blows over a soil surface which is dry and has no surface vegetation cover. The fine material is blown from between the pebbles so that they eventually form a complete surface cover, an armour against deflation.
A desert caused by such pebbles is called a reg.

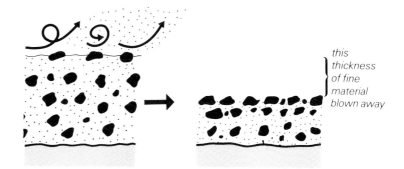

Exercise 16.3 Wind transport and deposition in the desert

1. Without a protective vegetation cover, soil is easily moved by both wind and water. Which size of sediment could be moved by wind action: boulders, pebbles, gravel, sand, silt, clay? There are several grades of material moved by wind. Particles of the largest moveable sizes are made into the features shown in Figure 16.7 and they form a sand sea. What are the irregular mounds of a sand sea usually called?

2. Silts are rarely found in a desert because they are blown away. However, in areas marginal to deserts they may settle out and accumulate to depths of over a metre. Silt deposits that have formed in this way are called loess. Sometimes when the wind blows from the southwest, summer rainstorms in New England wash a red dust — from the deserts of the southwestern USA — out of the air. It is deposited as a fine layer over cars, roads, and houses. What size of material is this and how did it manage to travel so far from the desert?

3. The shape of sand dunes usually reflects the direction of the dominant wind. Therefore, if they are regular in shape like **barchans** (Fig. 16.10), the wind must come predominantly from one direction. Describe the barchan's shape, make a sketch of it, and try to explain why its "horns" move ahead of the main dune.

4. Barchans are rare, isolated dunes. Long straight sand dunes, called **seif** dunes, are found only in the places where wind directions are constant. So what sort of wind pattern might have caused the irregular surface of the sand sea in Figure 16.7?

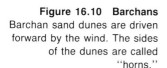

Figure 16.10 Barchans
Barchan sand dunes are driven forward by the wind. The sides of the dunes are called "horns."

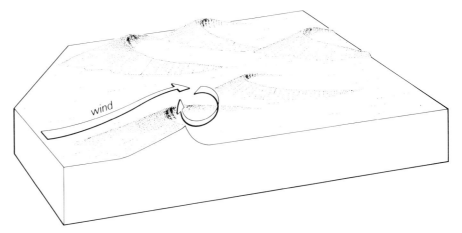

So, except in areas near to the ground, wind is more of a transporting than an eroding agent. *Very few landscape features have been produced by wind alone.* You must remember that both wind and water are at work in a desert. Water aids weathering and breaks the rock up into fragments which can then be carried away by the wind.

Wind transport is not confined to deserts. Wind will blow light, loose materials around on any part of the earth's surface where vegetation is sparse. Sand dunes and beaches around ocean coastlines are obvious places where wind action can be experienced, as anyone who has had a picnic on a beach knows. Near to glaciers the land is also bare and fine material is easily blown away. Indeed the fine material from both hot deserts and glacier margins has, in the past, been blown away and settled onto much of northern Europe, America, and China. Today it provides some of the world's most fertile soils (called **loess**).

Recently wind erosion has been given the chance to work in new areas. People have been busily removing vegetation and ploughing land for thousands of years. Over the past hundred years we have been very clever at preventing the return of the unwanted natural vegetation (weeds) by using herbicides. We have let large barren fields dry out in spring before plants can grow and protect the soil. These areas are ripe for wind erosion (Ch. 12). Soil is blown away from many areas in winter, but the most disastrous results occur when seedlings emerge, because wind erosion will then blow millions of sand-size soil particles across the fields and slowly shred seedling leaves by abrasion. Not surprisingly, this process, which leads to the expansion of deserts, is called **desertification.**

Islands of water, seas of sand   *Oases*

Despite its widespread effect, wind action rarely creates distinctive landforms. Abrasion mainly results in smoothing the base of cliffs or upstanding rocks. However, one feature in a desert might not exist but for the wind. In areas where water is near the surface, it can rot the rock and make it easy for wind to blow material away. This forms large, shallow depressions which eventually expose the water source. Shallow water-filled depressions in deserts are called **oases** (Fig. 16.11).

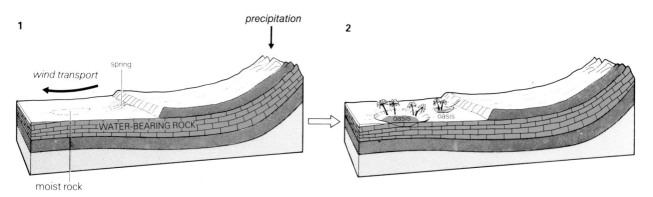

Figure 16.11 Oases (1) Water-bearing rock near the surface allows chemical weathering and the breakup of surface rock. (2) Wind removes loose, rotted rock and exposes water-bearing rock. The permanent pool thus provided allows vegetation and human occupation. Some oases stretch over many square kilometres.

17 Glacial landforms

Ice advance

A chilling tale

Chamonix, in the French Alps, developed as a small thriving market town during the Middle Ages. In the nearby fields the farmers grew barley and other crops. High above the valley in which Chamonix lay stood the majestic snow-capped peaks of the Alpes Maritimes (Fig. 17.1). It had always been like that as far back as anyone could remember. But from the year 1600 terror and death came to the townspeople of Chamonix as, season by season, the weather deteriorated until what had once been a green and fertile valley was invaded by a huge tongue of ice.

By the 1640s government officials, who had been sent to the area to investigate the low payment of taxes, reported:

> The glacier "Des Bois" advances by a musket shot every day, even in the month of August, towards the said land of Chamonix. We have also heard it said that there are evil spells at work among the said glaciers and that the people implored God's help to preserve them against the said peril. The people sow only oats and a little barley which, throughout most of the seasons of the year, is under snow, so that they do not get a full harvest in three years, and then the grain rots soon after. The people there are so badly fed they are dark and wretched and seem only half alive.

Soon this advance of ice into the once-fertile valleys of the Alps became "normal" and the area was abandoned. However, eventually the climate improved again, the glacier retreated, and today Chamonix is once again an agricultural market town. And the ice? It is back up in the mountains and once again just part of the backdrop to be admired by tourists.

Figure 17.1 The French Alps, Mont Blanc and Chamonix, France

Figure 17.2 Arctic Maritimes?
With just 1°C drop in average temperature this is what the highlands of the Maritimes might look like in summer!

Glacial advances or surges, like the one that destroyed Chamonix, can happen quite suddenly, within a few years of a small downward fluctuation in average temperature. Not even southern Canada is secure: a decrease in average temperature of just 1°C would bring glaciers back to the highlands of the Maritimes (Fig. 17.2). And if the decrease in temperature falls a mere 6°C on average, then we will be back in conditions like the "Ice Age," with ice sheets spreading over the Canadian Shield, the Great Lakes, and the St. Lawrence lowlands. Indeed ice in Canada began to retreat only about 15 000 years ago (Fig. 17.3). So, although a "greenhouse effect" seems to predominate now, the fortunes of Canada, and for that matter, Britain, northern Europe, and northern Asia hang in a very fragile balance.

Clearly the changes required to go from one climatic regime into another are small. In the past half a million years (the period often referred to as the "Ice Age"), there have been several advances and retreats of ice sheets. Between each **cold period** there have been warm periods (called **interglacials**) during which the temperatures were often higher than those we experience today. Some scientists think we are just in another interglacial period today.

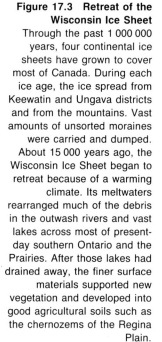

Figure 17.3 Retreat of the Wisconsin Ice Sheet
Through the past 1 000 000 years, four continental ice sheets have grown to cover most of Canada. During each ice age, the ice spread from Keewatin and Ungava districts and from the mountains. Vast amounts of unsorted moraines were carried and dumped. About 15 000 years ago, the Wisconsin Ice Sheet began to retreat because of a warming climate. Its meltwaters rearranged much of the debris in the outwash rivers and vast lakes across most of present-day southern Ontario and the Prairies. After those lakes had drained away, the finer surface materials supported new vegetation and developed into good agricultural soils such as the chernozems of the Regina Plain.

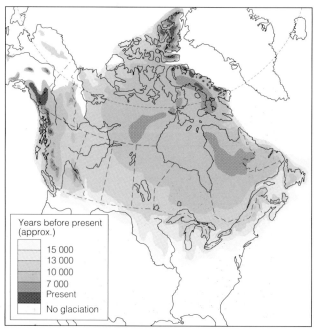

Records show that interglacials usually last between 8000 and 12 000 years, and it has been well over 10 000 years since the present interglacial began! It seems we can confidently expect colder times ahead, although the change is not likely during our lifetimes. Nevertheless, this "Little Ice Age" that sprang so suddenly upon the 17th century shows how quickly conditions can change.

Permanent snow

In search of snow

One of the newest ski resorts in North America is not in the Rockies or the Sierras, but in the southern Coast Range of British Columbia. Near the town of Whistler, the Pemberton Valley provides the mountains with a general northwesterly aspect (Ch. 5). Wedge Mountain (2890 m) dominates the impressive topography while providing local relief of over 2000 m. Several factors make this an attractive area for alpine sports, including its proximity to Vancouver and the Seattle-Tacoma area (Washington State). These factors are: altitude, aspect, latitude, exposure to orographic precipitation (from prevailing westerly winds), and the varied terrain of a maturely dissected region. But the drive to Whistler is not always easy. The building of roads and railways along the shores of Howe Sound and the valleys north of Squamish has proven to be exceptionally difficult. Also their maintenance in inclement weather is virtually impossible.

Some avid skiers look for fresh snow in the remote, rugged Purcell Mountains in southeastern British Columbia. For the past 20 years, helicopters have been taking skiers into remote areas of "The Bugaboos." Skiers might have 20 000 vertical m of skiing during a period of several days. Fortunately, heli-skiing guides are adept at finding vast fields of fresh snow and avoiding avalanches!

These examples show that snow survives because of a combination of *altitude, aspect* and *latitude,* with the permanent snow line being lower towards the North Pole. So when a cold period begins and snow and ice stretch out beyond their present limits, the areas most quickly affected are those in northerly latitudes. Even at the maximum spread of the ice during the Ice Age, glaciers from the southern Rockies, the Sierras, and the European Alps could never spread very far from their mountains. They were simply too near the Equator. When the ice spread over North America, Europe, and Asia, it came from the north, overwhelming everything in its path.

Exercise 17.1   The snow line

1   Draw a graph of altitude (vertical axis) against latitude (horizontal axis). Mark on the positions of the permanent snow line of the following mountains:

2   Join the plotted points with a smooth curve. What does the graph show? Wedge Mountain's height is 2890 m and its latitude is 50°N. Plot this on your graph. Do you think Wedge Mountain has a permanent snow cap?

Mountain	Snowline height (m)	Latitude
Mt. Kenya	5500	0
Diavolezza	3200	46
Svartisen	1500	67
Arctic pack ice	0	80

The change from snow to ice

The freeze

Snowflakes are among nature's most delicate objects (Fig. 17.4). When snow falls, millions of tiny flakes drift silently to the ground and settle into a light, easily blown mass. It can take a metre of newly fallen snow to make the equivalent of only 120 mm of rain.

When snow falls it clothes trees in a delicate cloak of sparkling white. But it also falls onto roads where it is compressed by the passing traffic; the delicate flakes are crushed within a fraction of a second. In this form, snow becomes brittle and hard. Road surfaces may become slippery and dangerous. Crush snow into a snowball and the same ice-making effect is achieved. Out on the mountainsides, the same process is at work but it takes months to crush snowflakes by the sheer weight of banked-up snow. Nevertheless, the snow eventually loses its delicate form, being transformed into small round crystals that are fixed together. Further compaction can take place quite quickly on a warm spring day. Warm air blows over the snow and transfers enough heat to melt the surface. The melted snow sinks into the gaps between the crushed flakes and then refreezes, helping to make the snow more and more like a solid mass of ice.

These two processes of crushing and refreezing gradually turn a hillside of snow into a brittle crystalline material that can lacerate the hand of a falling skier. It is called **névé** by the French and **firn** by the Germans. But it still doesn't make real ice; many years are required over which crushing and refreezing work while the snow gets thicker and thicker.

Ice behaves quite differently from snow. Snow is made from tiny independent flakes which have little combined strength; ice is a solid mass that has great strength both to support a weight and to tear and transport a rock (Fig. 17.5). When snow turns to ice then gentle hillside hollows can be eroded into deep bowl-shaped features, boulders ground to rock flour, and blocks torn from the solid rock.

Figure 17.4 Snowflakes
Snowflakes are created when water freezes in groups of 6 molecules. They form slim, star-shaped crystals which add complexity to their shape, size, and design by attracting more water molecules. It is the intricate shapes that allow snowflakes to lock together and hold snow at steep angles on mountain sides.

Figure 17.5 Glacial transportation
Moving ice often breaks large rocks from their bedrock origins and carries them great distances. This rock, near Drumheller, Alberta, is a well-known erratic which has been carried at least 500 km from the Canadian Shield. The ground near the erratic has been packed down by the hooves of bisons which have, over thousands of years, used this rock as a "rubbing post" in an otherwise featureless grassland area.

Figure 17.6 Under the ice *Above (a).* Ice is a solid mass that has strength enough to support a cave, but it also flows slowly to give the beautiful arch shown here. *Below (b).* As ice moves from the rock and arches out to form a cave, it "sticks" to the rock and pulls small pieces away. In effect the ice is **quarrying** the back of the rock step. This material is carried across the cave roof and shows as the dark smear moving from right to left. Pieces of rock quarried from the cave's back wall are partially frozen into the bottom surface of the ice. *Below right (c).* When the rocks reach the lower end of the cave, they are pressed into the ice and form part of the glacier's debris load called **ground moraine**. At the same time they abrade the rock to form grooves called **striations**.

Nature's stonemasons *Glacial erosion*

Some of the most intricate shapes created by people are in the stonework of cathedrals and Parliament buildings. They are the work of generations of patient effort by stonemasons. The tools of the stonemason are the mallet, chisel, and file. The mallet and chisel chip off pieces of the rock and leave the surface rough and uneven. The file, on the other hand, scratches the surface with many fine teeth and leaves a much smoother finish.

Glacier ice is nature's most effective stonemason. Ice is strong and it can carry blocks of rock on its undersurface across other rocks (Fig. 17.6b). This is the glacial file. The process is called glacial **abrasion**. Ice can also pluck at rock and rip it away from hillslope or valley floor. This is the mallet and chisel. The process is called **quarrying**. You can get an impression of the pulling force of moving ice when food packages often become stuck to the inner surface of a freezer, but it is best seen in an ice cave (Fig. 17.6a).

Figure 17.7 A roche moutonnée

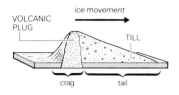

Figure 17.8 A crag and tail

Ice moves because of gravity. The mass of a hollow or valley full of ice will cause it to slide very slowly. All the time the ice is sticking to its floor, quarrying loose blocks away and then pressing them down hard and scraping them across the rock that remains in place. The results are very distinctive. Any projecting rock on the valley floor becomes ground smooth and covered with scratches on the side facing the ice movement. The scratches are called **striations** (Fig. 17.6c). On the downslope side, the rock is quarried and becomes jagged. The final sculpted rock is sometimes called a **roche moutonnée** (Fig. 17.7). It is not usually very large, being a few tens of metres long, and is just an irregularity in the bedrock. Larger obstructions effectively protect their downslope sides and may even be a place of deposition for glacial debris — as though the glacier had swept its chippings out of sight behind a rock (Fig. 17.8). This is called a **crag and tail**.

However, in the mountains erosion is dominant and glacier ice uses the abrading and quarrying tools to create the very special type of landscape.

Cirque formation

Giants' armchairs

The first places snow turns into ice at the start of a cold period are the headwater hollows that once fed rivers. These headwaters are high and so will be the first areas to experience freezing temperatures. Together, low temperatures and sheltered hollows allow snow to build up and be crushed

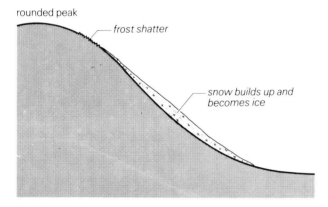

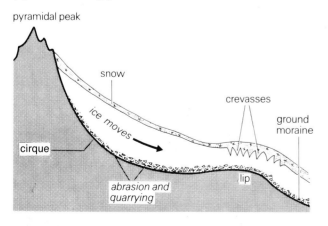

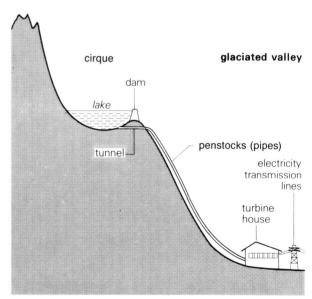

Figure 17.9 Cirques
(a, b) The sequence of change from valley head to cirque.
(c) The use of a cirque for hydro-electric power.

Figure 17.10 Ice in a cirque
A cirque with ice still eroding it, *above* from straight on and *below* from the side. Notice how the ice appears to have slumped down from the back. Crevasses show where the stresses occur in the ice except near the top where it is covered by snow. The photographs were taken in late summer when snow cover is at a minimum.

into ice. Slowly the ice slips down the hollow and down the hillslope towards the valley. As it does so, the ice mass rotates and scrapes at the rock so that, in time, the hollow changes shape to allow the more efficient movement of ice (Fig. 17.9). All the time the ice is moving out of the hollow, more snow falls and builds up into new ice. After thousands of years the result is to produce a bowl-shaped feature like a giant's armchair set in the mountain side. Ice-eroded hollows are very distinctive. They are called **corries** in Scotland and **cwms** in Wales, but most North Americans use the French name, **cirques.** They come in many different sizes and are hewn from all types of rock, but through all this their distinctive shape remains (Fig. 17.10). Indeed, they are probably the clearest evidence of the differences between erosion by water and by ice. Bowl shapes, where the

Figure 17.11 Ice in the landscape
(a) Before glaciation, (b) during glaciation, (c) after glaciers have melted

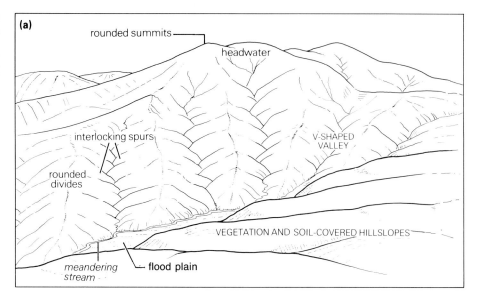

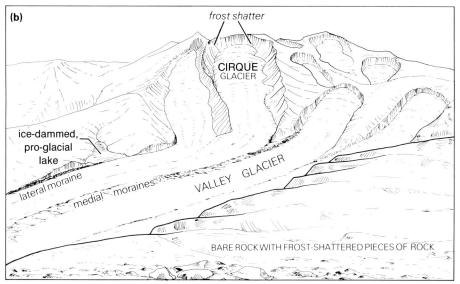

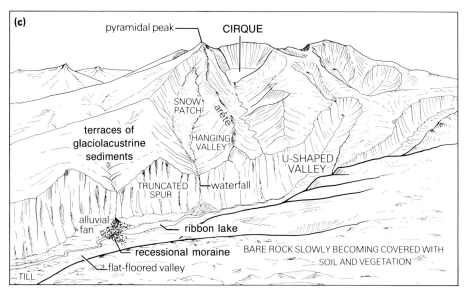

outlet **lip** is higher than the centre of the bowl, would not be possible with water erosion. This is why many cirques fill with water and form the high lakes (**tarns**) that make ideal sites for small hydro-electric power stations.

Before glacial conditions arrived (Fig. 17.11a), a mountain summit may well have formed the starting point for several streams and it will, therefore, be the site of several headwater hollows. Gradually, ice develops in each hollow and, as described above, enlarges them into cirques. At the same time the mountain peak will become more jagged. Indeed, cirques eventually compete for space around the peak, enlarging themselves until only knife-edged ridges called **arêtes** remain to divide one cirque from another. The Matterhorn in Switzerland is perhaps the most famous example of a glaciated mountain where four cirques have reduced the peak to a jagged pyramid. However, there are many other areas with **pyramidal peaks,** including those in Figure 17.13 and Mt. Robson in Canada.

Battle in the valleys

Glaciated valleys

Figure 17.12 Crevasses Ice breaks and produces deep crevasses when it is forced to change gradient quickly such as at an ice fall.

As ice slowly squeezes out over the lip of a cirque and down the hillside it is already carrying a vast supply of debris of all sizes. Now these weapons are used to do battle with the valley rock below.

When ice first enters a valley it moves into a landscape created by flowing water (Fig. 17.11a). It moves across hillslopes that were once covered with soil, grass and trees. But in the cold climate that allows ice to advance, the trees and grass have long since died and the exposed soil has been stripped off by avalanches and other mass movement processes. Now, only bare rock remains strewn with frost shattered debris. Nevertheless, the valley shape has to be changed until it is more effective at allowing ice movement. Water can meander and flow quickly in a narrow channel, but ice is slow moving and does not easily change shape to flow around bends. So while water once tumbled along the bottom of the valley, it now becomes locked up in a dirty grey creeping mass of ice hundreds of metres thick. (Fig. 17.11b).

Valley and cirque glaciers erode in similar ways, but in a valley the volume of ice is greater and the area of abrasion much larger. Here the task is to create a new valley shape which is more suitable for ice flow. It must be broad and deep and have no obstacles or tight bends. Abrasion and quarrying work fastest where such constrictions and obstacles exist, because in these places ice must flow faster and so scrape more debris over the valley floor. Erosion continues day and night, year in year out until *a new valley shape is formed with concave sides, a wide floor and reasonably straight plan.* No longer do hill spurs project across the valley. They have all been **truncated** (cut off) by ice erosion. The small streams of ice that flow in tributary valleys are not able to keep pace with the downward erosion of the glaciers in the main valleys and soon they are left as **hanging valleys** with icefalls cascading from them (Fig. 17.12). To achieve this, millions of tonnes of rock have been removed and carried away in the lower layers of the ice. Erosion by valley glaciers continues so long as new ice forms in cirques. But although the valley shapes have been modified, gravity still ensures that glaciers follow the old paths of the rivers to the sea.

Sometimes valley glaciers erode some parts of their valley floors deeper than elsewhere. This may be because the bedrock is less resistant, or the glacier might be more powerful after receiving ice from a tributary. Overdeepened sections of valleys may fill with water and become **ribbon (finger) lakes** when glaciers melt away (Fig. 14.22b-d). Shuswap Lake and Okanagan Lake are two such lakes.

Unless ice overwhelms a valley, slopes above the valley glacier will remain bare and vulnerable to frost shatter. As blocks are shattered from the slopes, they fall onto the edge of the ice. Accumulations of frost-shattered debris at the sides of a glacier are called **lateral moraines.** Where two glaciers join, the common lateral moraine will be carried between the glaciers as a **medial moraine**. Surface moraine carried away from the margin of the valley is called ground moraine. It is rare for any surface moraines to remain as distinct ridges after the valley glacier has melted.

The main supply of debris is carried in the lowest layers of the ice and is obtained by abrasion and quarrying. It is called **ground moraine** or **till.** When a glacier melts away, till covers the valley bottom and may well give the valley a flat floor. If the glacier has remained stationary in a valley for some time, a **terminal** or **recessional moraine** may form. When the ice melts, this may hold back some of the **meltwater** and create a lake (Fig. 14.22 b-d).

Exercise 17.2 Alpine-type glaciation

In some isolated high mountain regions such as the Alps, Himalayas, and Rockies, glaciers did not fill their valleys, even at the height of the cold period. As a result, mountain summits continued to be weathered by frost shatter over long periods and they became sharp and jagged.

1 Study Figure 17.13, which shows part of the Oberaar valley in the Swiss Alps. Notice that ice occupies only a small part of the valley today.

2 Match up the features labelled A, B, C, D, E, F, G, H with the following terms: glaciated valley, arête, frost-shattered peak, pyramidal peak, cirque, truncated spur, ice fall, and hanging valley.

3 Explain how truncated spurs and hanging valleys form.

Figure 17.13 The Oberaar glacier, Switzerland

Figure 17.14 Moraine Lake, Alberta
This small rock basin lake high in the Canadian Rockies is clear and colourful because of reflection of light from deposited rock flour from nearby glaciers. The most remarkable features are: the arêtes, pyramidal peaks, sedimentary rock strata, steep scree slopes, and U-shaped valleys. Above Moraine Lake, the arêtes form both the continental divide and the BC-Alberta border.

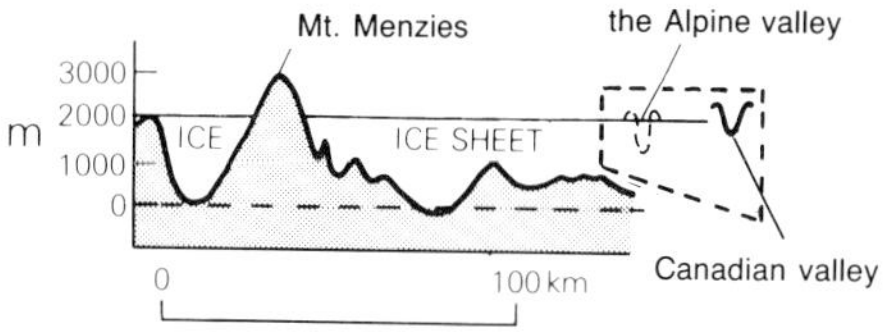

Figure 17.15 Antarctic ice
A cross profile of a small part of Antarctica showing peaks standing above the ice and large numbers of valleys engulfed by it. (The Alpine valley, Fig. 17.13, and the Canadian valley Fig. 17.14, have been drawn into this diagram to scale and in their correct relationship to the ice level.)

Exercise 17.3 Canadian glaciation

Most of Canada has been thoroughly glaciated during the past 20 000 years. The majority of eastern and central Canada was covered by massive, slow-moving ice sheets that originated in the Keewatin and Ungava districts. At the same time, the arctic islands experienced some valley glaciers, but were otherwise ice-free. The mountain tops of the Cordilleran Range were also ice-free. Alpine glaciers radiated from their cirques; the mountains weathered into pyramidal peaks.

1. List the visible features in Figures 17.13 and 17.14.
2. The arêtes shown above Moraine Lake (Fig. 17.14) are part of the North American continental watershed. What provincial boundary do they form?
3. Moraine Lake is a rock basin (ribbon) lake. How might it have been formed?
4. Figure 17.15 shows part of Antarctica. Most of the continent is under 2000 m of ice even today. If all the ice were to melt, what features would you expect to find on Mt. Menzies that would not be found on other mountains in the cross section?

SUMMARY: FEATURES OF MOUNTAIN GLACIATION

Snow first builds up in the sheltered north-facing hollows that were once the headwaters of streams. As snow accumulates it is crushed into **névé** and finally into **ice**. Ice within a hollow moves under its own weight, **quarrying** rock from the hollow and using it to **abrade** yet more rock. Eventually a **cirque** (corrie, cwm) forms with its characteristic bowl shape. Ice within a cirque is called a cirque glacier.

If cirques form side by side, the rock separating them may be eroded back until it is a sharp ridge. This is called an **arête**. Sometimes several cirques form around a mountain and erode it into a sharp pointed **pyramidal peak**. These peaks have arêtes radiating from their summits.

As ice from several cirques flows into a valley, the ice makes up a **valley glacier**. Valley glaciers abrade and quarry in just the same way as cirque glaciers do. The tell-tale scratch marks of glacial abrasion are long thin grooves in the rock called **striations**. Quarrying leaves rough jagged rock surfaces. Rocky remnants of valley floors that have been abraded on one side and quarried on the other are called **roche moutonnée**.

Prolonged quarrying and abrasion straighten out river valleys and produced a **U-shaped valley** with **truncated spurs**. Because main valley glaciers are more effective at erosion than their smaller tributaries, glaciers in tributary valleys are often left as hanging glaciers in **hanging valleys**. When the ice melts, the hanging glaciers are replaced by waterfalls. Uneven erosion leaves many over-deepened sections on the valley floor. After glaciation these will be filled with water and form **ribbon lakes** (Fig. 14.22b).

Some of the eroded debris will be left behind as **ground moraine** or **till**. **Lateral** and **medial moraines** that formed on the ice surface may remain as ridges on the till, but they are not as common as the eskers (Fig. 17.20) which mark the courses of ancient subglacial streams.

The limits of glaciation

A change of pace

Sooner or later glaciers reach the more gentle gradients of lowlands, and as they move onto lowlands their paths are no longer guided by valleys and they can spread out and coalesce to form large continuous areas called **ice sheets.**

Lowlands have gentler gradients and ice, therefore, moves more slowly, often pushed only by the mass of ice accumulating in upland valleys. Even so it can still perform much erosion, and evidence of this is widespread. For example, the ancient granitic rocks of the Canadian Shield have been eroded to a low relief and gently undulating topography by several ice sheets

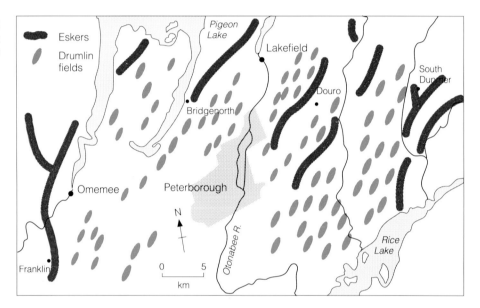

Figure 17.16 Drumlins and eskers of the Kawartha Region, Peterborough, Ontario

which originated in the Keewatin Region (NWT) and Ungava Peninsula (Quebec) during the past 500 000 years. The massive ice sheets scooped out a multitude of rock basin lakes, and scoured the granitic bedrock. Each ice sheet deposited another heavy layer of sediment on the Prairies as the ice slowly ground its way outwards from the centres of ice accumulation. At the ultimate extent of glacial activity, a terminal moraine was dropped as an irregular line of gravelly hills. Towards the end of each glacial period, a general warming of world climates reduced the supply of snow to the glaciers and slowed their advance while returning much of their bulk, as meltwater, to lakes, rivers, and oceans.

Although ice on lowlands can perform much erosion, the more it erodes, the more debris it has to carry. Carrying millions of tonnes of debris is an energy-consuming business, especially on a lowland where the energy supply is low. Finally a stage comes when the ice runs out of energy and erosion ceases. It is like a train full of passengers — it can move onward but it cannot pick up any more until it has dropped some off. Sometimes dropping zones are very localized and piles of till are deposited as hummocks called **drumlins** (Fig. 17.16), but most deposition has to wait for a climatic warming and ice melt. In the meantime, ice continues to spread until it is no longer pushed forward from behind.

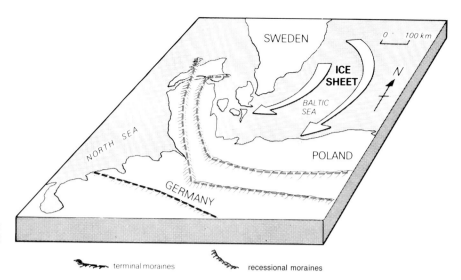

Figure 17.17 Deposits of an ice sheet

Exercise 17.4 Features of ice deposition

Till deposition has created most of Denmark. This also holds true for northern Germany, the Netherlands, the Prairies and the St. Lawrence lowlands.

1. What are the half egg-shaped mounds of debris that are deposited beneath ice sheets?
2. How and where do terminal moraines form?
3. Eastern Denmark is composed mostly of till while western Denmark is formed from sand and gravel outwash plains. How might these regions have formed? What distinctive feature separates them (Fig. 17.7)?

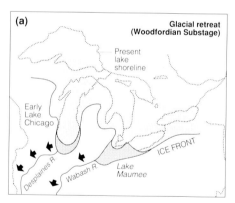

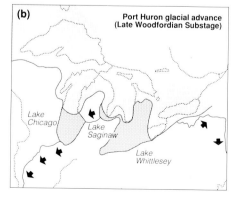

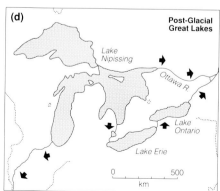

Figure 17.18 Probable evolution of the Great Lakes during the past 18 000 years
Since the last of the ice disappeared, isostatic uplift of this area redirected drainage through Lakes Erie and Ontario and changed the shape of the St. Lawrence estuary.

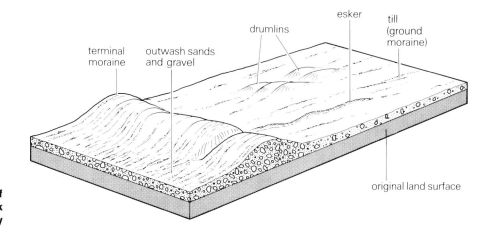

Figure 17.19 Features of glacial deposition in Denmark and north Germany

Figure 17.20 Esker at Norwood, Ontario
This long, sinuous ridge of poorly sorted moraine forms a 10 km esker at Norwood, near Peterborough, Ontario (see eastern edge of map, Fig. 17.17). This esker, one of ten in the Peterborough region, was deposited by irregular stream flow beneath a waning ice sheet towards the end of the Wisconsin glaciation.

Lowland glaciation **The melt**

Eventually the leading edge of the ice spreads so far from its source regions that it melts as fast as it is replaced by new ice from behind. When this happens the ice front becomes stationary, but it is far from inactive. Ice and rocks are still delivered to the melting front. Here ice melts, shedding its debris load which slowly accumulates as a ridge. This is a **terminal moraine** (Fig. 17.17). Some terminal moraines are major landscape features. Ridges nearly 100 m high, several kilometres wide and hundreds of kilometres long, occur in north Germany (Fig 17.19). The "backbone" of the Jutland peninsula of Denmark is a terminal moraine. Terminal moraines from the Wisconsin glaciation of North America appear as distinctive, uneven ridges running across the northern states of the USA from the Cascade Range in Washington State to the Appalachians in New York.

Meltwater streams do not begin at the ice margin but often tens of kilometres back within the ice. This tells us that warmer air causes ice to melt over a wide zone near the ice margin, not just near the ice front. As a result of melting, streams form on the ice surface but quickly disappear down crevasses until they are flowing between ice and bedrock. As they flow they erode tunnels in the ice and also pick up much till and carry it towards the ice margin. Sometimes the beds of such streams can still be recognized as winding ridges of forested gravel. These ridges are called **eskers.** Few eskers have been preserved in Canada, except around Peterborough, Ontario (Fig. 17.16). They are still well preserved behind the terminal moraine in the northern USA, in Ireland, and in Finland they are very common and have dammed up water to form famous lakeland regions.

Meltwater streams bring much debris to the ice margin and they carry some of it away. No longer confined to ice tunnels with efficient channels, they spread out into braided channels in front of the ice where they deposit the pebbles and gravel in a wide zone called an **outwash plain.** Like till, outwash plains are very extensive, but they do not make distinctive landscape features. The gravels have since weathered to produce only infertile soils and are mostly hidden under forests.

SUMMARY: DEPOSITS OF GLACIERS AND ICE SHEETS

Glacial deposits are of two types: (a) angular scree that falls on the ice surface from nearby frost-shattered slopes and (b) rounded material that has been quarried from beneath the ice.

Scree materials carried along by a glacier on its surface are called **lateral moraines.** The inner stripes are made from the same material and they also rest on the glacier surface. They are called **medial moraines** (*medial* means middle). Medial moraines are formed by two lateral moraines when glaciers meet (Fig. 17.11).

The majority of the material carried by a glacier is in the bottom ice layers and is called **ground moraine** or **till.** On lowlands till often masks the preglacial landscape to a depth of tens of metres. Sometimes it forms streamlined mounds called **drumlins.** (Fig. 17.16). Most often till is deposited haphazardly over the landscape, filling in some valleys and making plains into areas of gently rolling hills.

At the limit of ice advance, a **terminal moraine** sometimes accumulates. This is a place where ice melts as fast as it arrives. Melting water carries the debris away in braided streams, gradually sorting fine from coarse over an **outwash plain.** Meltwater from behind the ice margin flows down crevasses and then runs in tunnels along the base of the ice, carrying washed out till with it. The winding ridges deposited by these streams are called **eskers** (Fig. 17.20).

Glacial melting ## The end?

As a cold period comes to an end and the climate warms, ice sheets seem to melt quite rapidly. Britain's ice sheets disappeared from the lowlands in a couple of thousand years. They vanished from North America within about eight thousand years. But their retreat into the mountains was slower. Many stopped in their tracks several times and deposited recessional moraines. Moraines still block valleys in southern British Columbia, but most have been carried away by rivers. For example, Penticton, British Columbia, is located on rearranged recessional moraines, glacial lake terraces, and alluvial fans which form a wide, low dam at the southern end of Okanagan Lake. But the ice does not give up easily. It has made a comeback three times in the past half a million years and it recently showed some signs of trying to do the same in the near future. In the meantime we can admire the shrunken remnants of ice sheets that we now call spectacular alpine glaciers and wonder about the time when ice spread over all of Canada to a depth of 2000 m. Ice left Canada and other northern continents less than 10 000 years ago. If it returns much of the Northern Hemisphere will look like Antarctica does today (Fig. 17.15). But where will the people live?

Past glaciations and people ## Glacial gifts

Glaciers are natural features that do not have much effect on our day-to-day life. Most glaciers are up in mountain areas, and any change in their size affects only the small number of people who live near them. Today people go towards glaciers; they do not run from them. Glaciers are tourist attractions, and each year thousands of people are taken for guided tours over them.

Ice sheets are rare. Small ice sheets occur in Norway, and somewhat larger sheets in Iceland and Alaska. The only two areas where really large ice sheets exist on land are Greenland and Antarctica — both places almost uninhabited.

But despite the remoteness of ice sheets today, the impact of past cold periods has been substantial. The alpine glaciers of western Canada grew out of the complex mountains of the Cordillera. The ice spread southwards, achieving depths of over 2 km, while carving out the deep trenches and mountain passes that now facilitate settlement, farming, and travel. The Okanagan Valley and Rogers Pass are two results of those glaciers. Highways resulting from ice erosion *across* high mountains are important in the Alps and North America as well. At the same time deep glacially eroded valleys with tributaries left hanging above them have also provided sites for hydro-electric power stations.

Probably the greatest legacy of glaciation is the least spectacular and least well known. Glacial deposition formed whole countries, such as Denmark, and substantial parts of many others. The Canadian Prairies have formed from sedimentation, the accumulation of hundreds of millions of years. The deep, dark chernozem soils are largely a gift of the Wisconsin glaciation and subsequent processes.

18 Coastal landforms

To catch a wave *Wave formation*

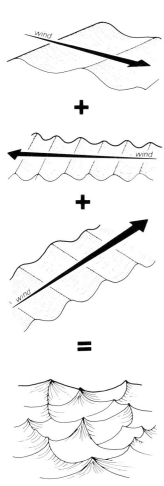

Figure 18.1 Ocean waves Waves in the ocean appear random and without pattern, but the peaks and troughs are the result of quite simple wave patterns that have been formed by winds far out at sea. Each pattern becomes superimposed on the others to produce an apparently inexplicable complexity.

When storm waves hit a cliff the pounding it receives is comparable to that of a nail being struck by a huge hammer again and again, ten times a minute, 24 hours a day. If they are to resist such forces most cliffs have to be very strong. People have often underestimated the power of the waves. At Wick, in Scotland, the end of a breakwater wall was built using an 800 t block of concrete secured to the wall with steel rods 10 cm in diameter. Yet soon after it was finished a storm wave threw the block and half of the wall back into the harbour which it was designed to protect. Together, wall and block weighed 1350 t. So the engineers tried again, this time with a block weighing 2600 t. Imagine their surprise when, a few storms later, this too was thrown into the water. In the USA there is a lighthouse beacon 45 m above sea level which has had so many rocks thrown through its windows by storm waves that a steel grating had to be built to protect it. And then there are storm waves lashed on by hurricanes. There are many reports of boats being carried several kilometres inland by waves tens of metres high. These are examples of **storm waves.** They are quite different from the **swell waves** that lap the beaches on a sunny summer's afternoon. But any wave can soon be whipped up by gale-force winds. A few hours of a strong wind blowing over an extensive ocean surface can turn waves a few centimetres high into gale-lashed giants the height of a house (p. 70).

Out in the ocean, water is always on the move — an ever-changing pattern of small wave crests and troughs (Fig. 18.1). As the wind builds up, it catches these small waves in the same way as it catches the sails of a yacht. As a result the waves are pushed across the ocean, building ever higher as they trap more and more wind.

However, waves are not just surface features. The water movements on the surface are tied to movements below, so that water is stirred up to a depth equal to about half the distance between wave crests. Usually this is nothing more than a surface ripple on a deep ocean, but as the waves approach shallow seas the stirred-up water starts to churn up sediment on the seabed. As soon as this happens, frictional drag with the seabed slows the waves down. At the same time they change shape, first becoming higher and then forming the **breakers** with which we are so familiar. Now as the waves complete their journey, all the power transferred from the wind is suddenly thrust upon the shore. The results can be dramatic, but they depend on the nature of the shore: there is a big contrast between the effects on cliffs and on sandy beaches.

Figure 18.2 Wave action on cliffs
As waves come onshore they increase in height and form breakers which then crash against cliffs. The pressure effect of such action can be dramatic.

Figure 18.3 Wave action on cliffs
(1) A wave breaks against a cliff made from jointed rock. (2) The wave breaks and water is thrust into the gaps between blocks, gradually dislodging surface rock. (3) As the wave retreats the loose block is pulled out and it falls to the bottom of the cliff, leaving more rock exposed to the waves.

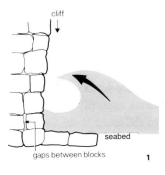

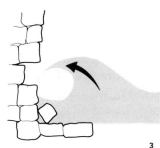

Crumbling cliffs *Wave erosion*

When a wave crashes against a cliff, all the pent-up energy is spent in a fraction of a second, yet the cliff yields but little. Instead, a massive curtain of spray is flung up over the cliff face as the water rebounds back into the sea (Fig. 18.2). However, cliffs are mostly made from jointed rocks, and the impact of the waves on joints is very different. Air is often trapped in the joints by the inrush of water and the water and air are driven like a hammer even deeper into the cliff (Fig. 18.3). The force of the wave is spent very quickly and, almost as soon as it has struck the cliff, water pours out of the joints and back into the sea. The sea is relentless and can repeat its hammer blows for as long as required. And patience is rewarded. Every so often a block is dislodged and the water is presented with a new piece of rock to attack.

The sea has other tools besides the force of water. Continual wetting of rock causes rotting (chemical weathering) in just the same way as it does on land. A few pebbles thrown against the cliff by the water every time a wave breaks have an abrasive effect very similar to sandpaper and every bit as effective in erosion.

Figure 18.4 Cliff erosion
The process shown in Figure 18.3 leads to undercutting of the cliff base. (1) With a small overhang the cliff remains stable. (2) Undercutting has now proceeded so far that the cliff is ready to collapse. (3) The cliff has collapsed and the resultant rock debris must now be removed.

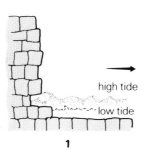
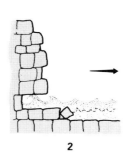
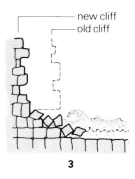

Cliffs are attacked with all of these tools between high and low tide level and the result of this concentrated attack is **undercutting**. An undercut cliff rapidly becomes unstable because of the large weight of unsupported rock above the overhang. As a result, cliffs crumble, crash, or collapse into the sea (Fig. 18.4). Whatever the speed of collapse, the debris scattered at the foot of the cliff protects the new cliff for a while. Now the waves have the task of breaking down the debris into pieces of moveable size. But as soon as the loose rock has been neatly cleared away, the waves resume their ceaseless pounding.

Coastline under attack

Caves, arches, and stacks

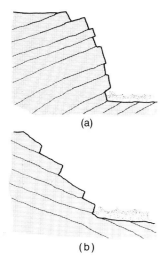

Well-jointed cliffs retreat by block removal and the cliff retreats in steps (Fig. 18.4). Notice that, in many ways, cliffs are very like hillsides. The angle of the cliff depends on the type of rock in which it is formed, the rate of erosion at its base and, sometimes, on the dip of the rock (Fig. 18.5). However, very few rocks have joints of just one size evenly spaced across a cliff face. More commonly, rocks have concentrations of joints in some places, fewer well cemented particles somewhere else, and perhaps a fault or two. As a result, cliffs are attacked most effectively in places, where the weaknesses are greatest, and sections of cliff retreat at different rates.

The first sign of weakness may be localized undercutting and the formation of a **cave** (Fig. 18.6). A stretch of coast with many weaknesses may be full of small caves and collapsing cliffs. Where there are many such weaknesses the coast will retreat rapidly and a **bay** will begin to form. Any resistant stretch of coast will be eroded more slowly and will gradually form a **headland**. Bays, inlets, and headlands are common Canadian coastal features. Notre Dame Bay, Newfoundland, has an espacially rugged coastline because of fault and rock types (Fig. 18.7).

Figure 18.5 Cliff profiles
The dip of rocks can affect the cliff steepness. (a) Steep cliffs with rocks that dip away from the coast. (b) More gentle cliffs with rocks that dip towards the sea. Blocks easily slide from the cliff in this case.

Figure 18.6 Summary of the stages of cliff retreat
(1) Caves are produced as waves erode the base of the cliff. (2) Collapse of cliffs above caves leads to cliff retreat and bays and headlands are produced. (3) Headlands are attacked from the sides. Through-caves produce arches; later, cave roofs collapse to produce stacks. The retreat of the cliffs leaves a wave-cut platform behind.

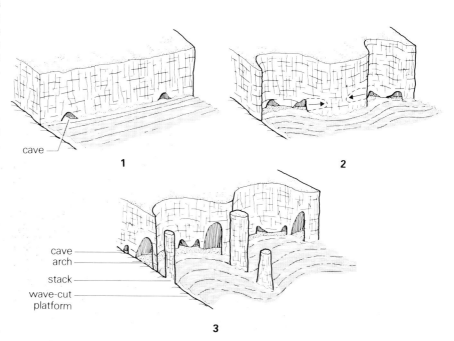

Figure 18.7 Inlets, bays and headlands

Twillingsgate and New World islands, Notre Dame Bay, Newfoundland, have particularly rugged, broken coastlines. The topography is a function of the complex geology, multiple faults, glacial erosion, inundation by the ocean, and the impact of storm waves which attack this area from the northeast. Notice that many small islands and stacks formed from conglomerate rock to the northeast of New World Island.

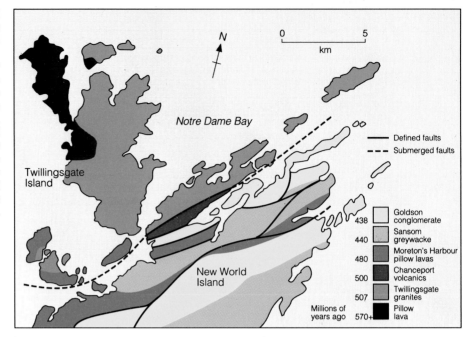

As soon as a headland is produced, it is attacked from the sides as well as head on. Enlargement of joints by waves may now lead to caves developing in the sides of the headland. If this happens and the caves meet, a tunnel is produced which, when enlarged further, leaves an **arch** which might be dry at low tide (Fig. 18.8). The slow demolition of King Arthur's castle at Tintagel (Cornwall, southern England) shows this well (Fig. 18.9). Rapid weathering soon caused such arches to collapse, leaving a pillar of rock (called a **stack**) or a rocky island to indicate the former position of the retreating headland (Fig. 18.7).

Figure 18.8 Hopewell Rocks, New Brunswick

Figure 18.9 The disappearing castle of King Arthur, Tintagel at Cornwall in southern England

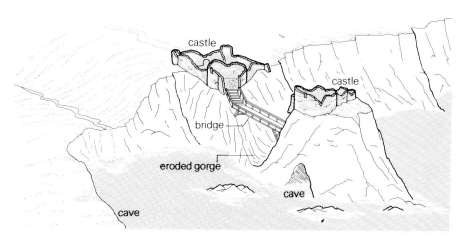

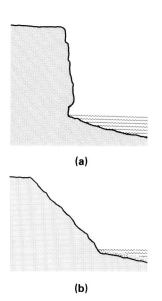

Figure 18.10 Cliff profiles

Exercise 18.1 Cliffs

Bays often have sandy beaches backed by gently sloping hills. When a wave breaks on a beach the energy given to it by the wind causes the sand and gravel to move. This uses up much of the wave energy.

1. What happens when a beach is not present to absorb the wave energy along a coast, say at a headland?
2. Figure 18.2 shows a wave breaking on a headland. Describe what is happening within the cliff. What is the result of prolonged wave action?
3. Match the two profiles of Figure 18.10 with the bay and headland of Figure 18.7a. Make a copy of these profiles and write "bay" and "headland" beneath the appropriate diagrams. Explain why you made your decision.
4. Figure 18.11 shows damaged oceanfront houses at Surfers' Paradise, Queensland, Australia. In June 1967, a rare winter tropical cyclone caused massive plunging waves to attack the open, quartz sand coastline. About 50 m of sand dunes were removed from in front of houses. A bulldozer was then used to replace loose sand in front of these houses at low tide. Why was this a futile activity? Suggest better ways of dealing with this problem.

However, wave attack causes little erosion below low tide and cliff retreat leaves a wide rocky **wave-cut platform** behind. Stacks and arches usually project from wave-cut platforms. These platforms are often very extensive and, on steep coastlines beneath a metre or two of beach, a wave-cut platform *always* hides.

Moving sand on beaches

Figure 18.11 Surfers' Paradise, Australia

Beside the seaside

Stand on any steeply sloping beach on a fine day and feel the water surging past your feet, pushing toward the land. Look down at the brown-coloured water of this **swash** containing thousands of grains of sand on the move. A few seconds later feel the water pouring back towards the sea, making wakes around your heels. Notice how the sand is pulled from beneath your heels by this **backwash.** Stand on a very steep beach made from pebbles and you can also hear the rush and watch the roll and tumble of the pebbles as they are carried back and forth with every wave. Stand on a very flat beach at the back of a bay and the waves come round your feet as thin sheets of placid water, each riding over the back of the one in front, but rarely appearing to return to the sea.

Do not stand on a beach in a storm. The rush of water would knock you off your feet and easily carry you out to sea. Each onrush of waves surges churning water over the beach and throws salt spray high up into

Exercise 18.2 Coastal erosion

Newfoundland's rugged coastline has been shaped by complex geological and geomorphological processes. Ancient granitic intrusions are exposed in several areas, but are usually covered by lavas — mostly basalt and andesite.

The lavas are mostly pillow formations which indicate that they erupted beneath the sea. The volcanic rocks were formed about 450 million years ago, when fish appeared as the first vertebrate animals.

Glacial erosion has shaped the hills and eroded many small fjords as ice repeatedly spread in a northeast direction from the interior of this large island.

1. Trace an outline map of Newfoundland from your atlas. (Select a scale of about 1:5 000 000.) Show scale and at least two parallels and two meridians.
2. On your map, clearly identify Notre Dame Bay and the area covered by the topographic map.
3. Identify on your map of Newfoundland at least six features that indicate the southwest to northeast trend of the island's topography.
4. Trace the coastline shown in Figure 18.12. Then, colour the inlets which run roughly southwest to northeast (blue) and at least six small hills which show the same orientation (red).
5. In one paragraph, explain why the coastline at Notre Dame Bay appears to have been flooded by the ocean, yet most of the land is composed of ancient rocks which erupted beneath the ocean.

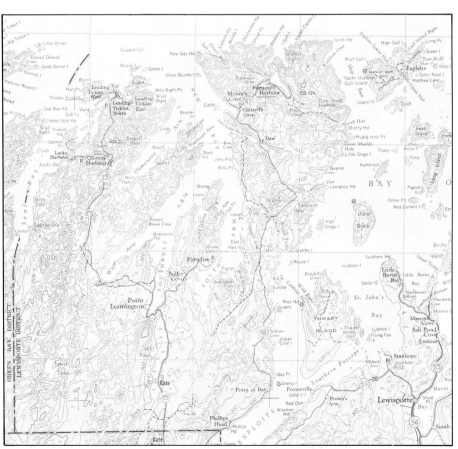

Figure 18.12 Part of the southern shore of Notre Dame Bay, Newfoundland

Scale 1:350 000

the air. Each curling wave thunders its way up the beach and hurls sand and pebbles before it. Only at the back of a bay are the waves gentle and calm.

Beaches are clearly places of considerable contrast. Waves change shape and behave differently because of variations in wave energy, beach slope, and beach sediment size. The same waves out at sea can behave very differ-

SUMMARY: COASTAL EROSION

Cliffs are eroded by (a) the *force of waves* thrusting water and air into joints in the rock (b) *abrasion* by pebbles thrown against the cliff, and (c) *chemical weathering* and solution.

Jointed cliffs retreat mainly by block removal. The sequence of steps is (1) wave action dislodges blocks from the cliff causing an overhang, (2) the cliff collapses, (3) the accumulated blocks protect the cliff and are slowly broken down by the waves and removed, and (4) this leaves the cliff exposed once more.

Because coasts are made from a variety of rocks, the least resistant will be removed first, possibly by the formation of **caves** and the collapse of **undercut cliffs.** Slowly, bays form which leave **headlands** of more resistant rock open to attack from the sides. Cave formation in the sides of a headland can cause **tunnels** and **arches** to form. Eventually the headland's earlier positions are marked only by lines of **stacks.** The width of cliff retreat is shown by the width of the **wave-cut platform** (Fig. 18.13). Many wave-cut platforms are not very wide because the sea has not remained at a constant level for very long.

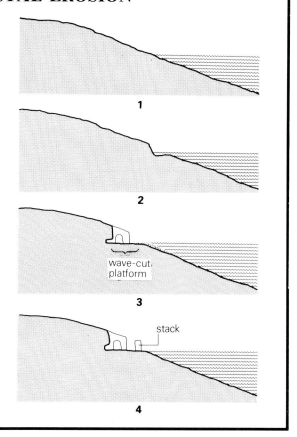

Figure 18.13 Stages in cliff formation

ently on a beach with a gentle slope and on one with a steep slope (Fig. 18.14). Material of pebble size moves quite differently from sand grains. These variations are the keys to the wide range of beaches that we see.

However, it is not just material in the surf zone that is on the move when waves break. The whole beach is being churned over to an area well outside where the breakers first start to form. And if the churning effect is noticeable on the edge of the surf zone, think how much more power is available to break loose and transport sediment beneath the place where the waves first break. We need to look closely at the form of the breaking wave at this critical point. Is it thrusting forwards and pushing material up the beach or is it plunging downwards and eroding the sediment from where it strikes?

Beaches are made from individual grains of rock which are easily moved by each breaking wave. This is the reason a beach never looks exactly the same from one day or season to the next. But the effect of the waves on a beach is very closely involved with the tides. While the waves are breaking, the tide is rising or falling, moving the shoreline in or out. **Spilling waves** that push material landwards will be helped by a rising tide and will leave material at the top of the beach, building it up. These gentle breakers are often called constructive waves. **Plunging waves** are erosive and they carry material seawards. If these waves occur on a falling tide, a metre's depth of sand can be stripped from a beach in just a few hours. These waves are often called destructive waves.

Topic Wave shape

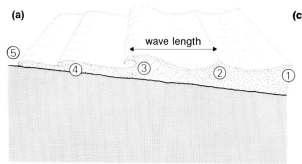

Figure 18.14 Waves on the beach
(a) Swell waves, (b) storm waves, (c) gently sloping beach, (d) steeply sloping beach.

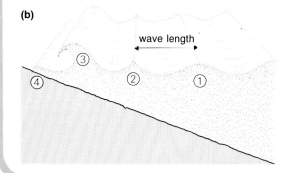

(a) **Gently sloping beach.** (1) Waves are affected by the sea floor some distance from the coast and they change from smooth rolling forms. The wave first develops a crest (2) which starts to break. As the waves move further inshore the crest water *spills* forward (3) as a breaking wave. The wave then rushes up the beach as "swash" (4) finally reaching the upper beach as a shallow flat sheet of water. Much of this water may sink into the beach sand so that little returns as a backwash (5).

(b) **Steeply sloping beach.** (1) Waves are affected by the sea floor much closer to the shore. All the changes of shape happen closer in. Here the wave becomes crested at (2) but the steep slope causes water to be pushed up the beach and the wave *grows* in height as it breaks. These waves *plunge* vertically and erode the beach. Such waves occur mainly during a storm. The steep beach prevents much forward rush of water (swash). In fact it encourages a backwash and helps erosion. Small waves on steep beaches do not become high enough to plunge and they tend to spill forwards, helping to repair any storm erosion. Waves that spill forward are sometimes called **constructive waves** (a + c); those with a large backwash and plunging breakers are sometimes called **destructive waves** (b + d).

The size of a wave, its energy, and its capacity to move sand onshore or offshore are all related. Large waves are produced by strong winds and are associated with storms. In many cases storm waves are plunging destructive waves (Fig. 18.14b). But storm waves at the back of a bay are rarely plunging, because the beach is too flat for them to develop; so destructive waves are mostly found on steeply sloping exposed beaches. By contrast, for most of the time spilling waves occur on all beaches. They are not driven on by any strong local storm wind and they are called swell waves (Fig. 18.14a). Their effect is much more subtle than that of the stormy giants. Because they have less energy, they move only small amounts of sediment forward at each tide, so it may take a couple of months to replace the material lost in a single tide of storm waves.

Figure 18.15 Stages forming longshore drift

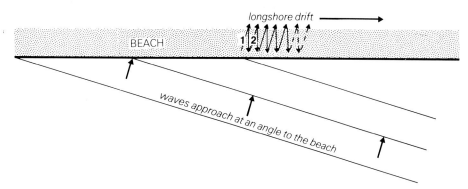

Longshore drift

Drifting sands

On an open coast, waves rarely approach parallel to the shore but arrive in long sweeping curves. There are no sharp bends in the plan of waves, and many coastal features reflect their shape. Bays, for instance, contain sand that forms long arcs of beach, widest at the centre, narrowest towards the headlands. Some sand escapes, however, and there is a gradual movement of sand (and sometimes even large particles such as gravel) along the coast. This is why it is very common to find pebbles on a beach that do not look like the rock of nearby cliffs. The movement of material along a coast is called **longshore drift**.

When waves reach the coast they usually approach at an angle. As the waves break they push material ahead of them in the swash, again at an angle to the coast (Fig. 18.15(1)). However, although the swash direction is controlled by the wave approach, the backwash is always down the steepest line of the beach (Fig. 18.15(2)). As a result, the beach material is moved along in a zig-zag motion called longshore drift. Because of this millions of tonnes of material are constantly moving along the beaches of the world every day. It is one of the main ways in which material is removed from the foot of cliffs, allowing further erosion to proceed (Fig.18.16).

Figure 18.16 Longshore drift (a) As waves break, material in the surf zone is pushed up the beach at an angle. (b) As the surf rushes back, it follows the line of steepest slope. The constant up and down motion of the beach material not only makes it rounded, but it also wears it into small fragments which can be more easily transported.

(a) swash

(b) backwash

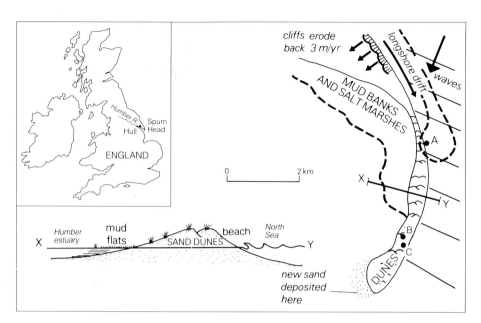

Figure 18.17 Spurn Head, a sand spit at the mouth of the River Humber

Spit formation

Sands of time

The drift of sand around coasts is a complex affair involving coastal current and the action of waves. But there can be no question that large amounts of sand, silt, and clay are on the move. Nowhere is this clearer than at the mouth of the Humber on England's North Sea coast. Here we find a long, irregular, curving ridge of sand called Spurn Head (Fig. 18.17). There is no rock core to this ridge even though it extends halfway across the river estuary. This long ridge of sand is capped by sand dunes which have been colonized by marram grass, and on its more sheltered river side, mud flats with salt marsh vegetation survive. It clearly receives its material from the north, where the coast is eroding rapidly.

We know much about the way Spurn Head has grown because it has been such a major hazard to navigation into the River Humber estuary and has been charted for hundreds of years. Imagine yourself in a small sailing barge in the 17th century. It is night and no light shows anywhere. There is a strong wind blowing and it is difficult to keep the boat running ahead of the wind. You are nearly home but you are straining your eyes for a sight of the low, dark silhouette of the dreaded sand bar that juts out in the estuary. Then, suddenly, the boat rears up and a grinding sound tells you the bottom is aground on the sand — you have struck Spurn Head and are now at the mercy of the pounding breakers.

Many deaths resulted from this unmarked **sand spit** and so it was decided to build a lighthouse at the end (A on Fig. 18.17). Soon after this lighthouse had been built the ease with which sand can be moved became apparent. The lighthouse was built on the edge of the spit, but as the years went by the spit grew longer and longer while the coast (and the spit) retreated further and further west. Finally 100 years after it had been built, the lighthouse was washed away. Clearly a new one was needed, so in 1776 it was built on what had then become the new tip (B). You can see that subsequently the spit has grown even longer and now even the new 20th century lighthouse (C) is not at the end.

Changes in sea level

The disappearing sea

Figure 18.18 Changing coastlines
(a) The coastlines of Alaska and NE USSR changed dramatically during the ice ages when sea levels were 100 m lower. A land bridge enabled people to trek across the Bering Strait. (b) If the ice locked up on land melted, Britain and Ireland would lose most of their lowlands and become an archipelago.

During the maximum extent of the last Ice Age, Nigalik and his extended family stood on a wind-swept cliff and gazed apprehensively to the east. To the south, the low, weak midday sun appeared briefly beneath a bank of heavy grey cloud. An expanse of frozen lowlands, which appeared to extend to the distant mountains, was visible. The ocean was more than a long day's trek towards the sun, but they would walk to the east. The population pressure and scarcity of game in their traditional hunting grounds had brought Nigalik and his family to the eastern tip of Asia. They knew they could not survive another long, cold winter. They would have to cross the wide, low land in search of food — across an unknown distance to a new continent.

It was an arduous, three-day journey, across what we now call the Bering Strait, to the new land. Nigalik's family may have been among the first people to cross the temporary land bridge between Asia and North America. Nigalik and his family struggled to survive because the ice that locked most of the northern lands in its grip 18 000 years ago supported little plant or animal life. The meltwater from glaciers and ice sheets provided little

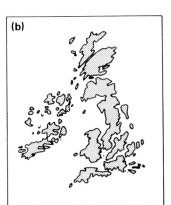

Exercise 18.3 Longshore drift

1. Because of longshore drift, sediment is moved continuously along coastlines. Figure 18.19 shows a bay and headlands. What has happened following the construction of a stone pier 400 m out to sea, thereby interrupting the flow of material?
2. Why would you use the northeast side of the pier for mooring boats, rather than the southwest side? **Groynes** are designed to prevent beaches from being washed away by longshore drift. They are needed where, for some reason, the normal movement of beach material along the shore is interrupted so that sand is lost and not replaced. The bay in Figure 18.19 did not need groynes before the pier was built.
3. Copy Figure 18.19 and draw on the pattern of sand you think would be trapped by the groynes (use the pier as a guide).
4. Why should anyone be bothered about beach loss?

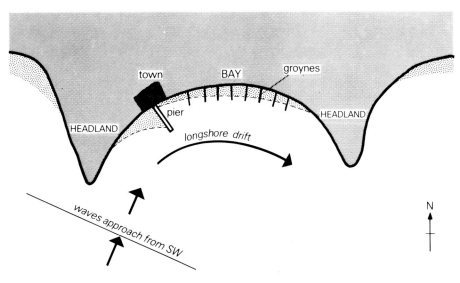

Figure 18.19 A sandy beach altered by construction of a pier

Figure 18.20 Drowned valleys

(a) Estuaries, (b) rias, and (c) fjords all represent coastal valleys deepened during the Ice Age that have subsequently been flooded. Many such as the fjord (bottom) at Pangnirtung, NWT, are now refilled with sediment, including scree and moraine.

nutrition for the sea life. But coastal rivers supported salmon in the summer, and the pack ice was home to seals and walruses in the winter.

Nigalik's family found a rich environment in North America. They and their descendants created a complex culture on the fjord coast of southern British Columbia. However, it took 2000 years before a noticeable warming of the northern climates caused a significant retreat of the ice sheets. Sea levels rose by almost 100 m and resulted in the flooding of the Bering Strait and the isolation of the populations of Asia and the Americas. Around the world, great coastal river valleys were flooded to create **estuaries** such as the Rio de la Plata (Argentina/Uruguay) and rias such as Chesapeake Bay (Maryland, USA) (Fig. 18.20).

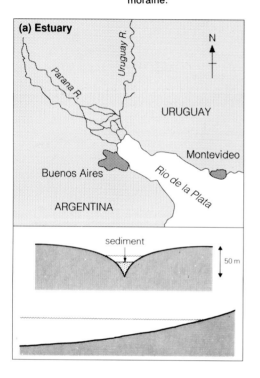
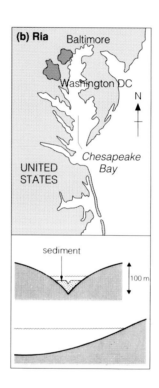
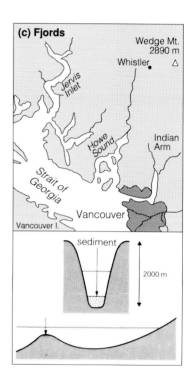

SUMMARY: COASTAL DEPOSITION

Some of the most interesting coastal features are associated with sand movements along the shore. Waves causing longshore drift of sand do not readily bend to take in sharp changes of coastline because they are controlled by the smooth shape of the seabed, *not* the coast of headlands and bays that form the edge of the land. So waves continue to sweep across the inlets of river valleys or bays until they are deflected by the force of river water, or they enter water too deep to cause breakers.

Because sand movement is related to breaking wave patterns, sand is dumped as soon as the water of the bay or inlet becomes too deep. Gradually the pile of dumped sand becomes large enough to reach the surface and form an extension to the land. This promontory of sand is called a **spit** (Fig. 18.17). The growth of a spit is controlled by a ready supply of sand, so spits are more common in places where the coast is easily eroded to provide the required material. Wind-driven waves from the southern Pacific Ocean sweep eroded material from the cliffs of southeast South Australia to form Younghusband Peninsula. This 150 km spit encloses a lagoon called The Coorong. Five low dams prevent salt water entering Lake Alexandrina, the mouth of Australia's largest river, the Murray.

More rarely, ridges of sand connect an island to the coast. Such a ridge is called a **tombolo**. A well-known double tombolo in Atlantic Canada connects Cape Jourimain to Cape Tormentine, about 3 km northwest of the New Brunswick-Prince Edward Island ferry terminal. This double tombolo encloses an island within its 4 km lagoon.

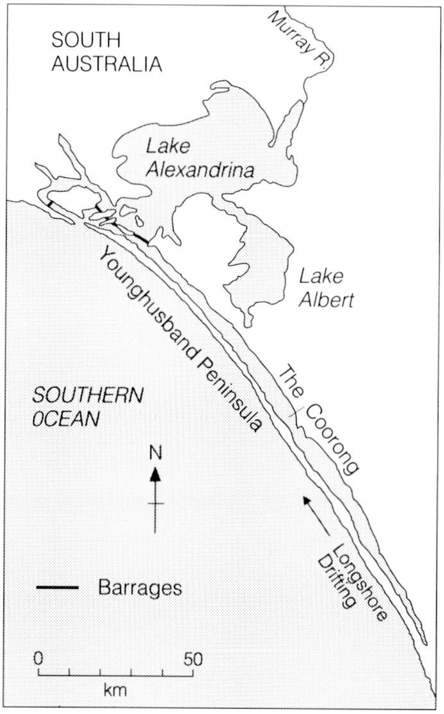

Figure 18.21 Younghusband Peninsula and The Coorong, South Australia
Longshore drifting has created one of the world's longest spits at the mouth of the Murray River.

Figure 18.22 Cape Jourimain, New Brunswick
Sediments deposited in the sheltered waters behind Cape Jourimain have created a lagoon enclosed by a double tombolo.

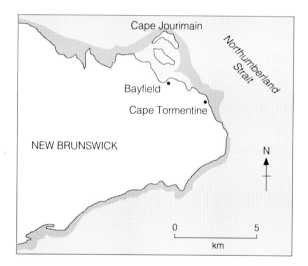

Finally, it is worth noting the effect that the sea can have on the coast if waves nibble at the edges of the land for a long period without a change of sea level. We have already seen that wave-cut platforms will replace cliffs. You may have already noticed how flat cliff tops often seem to be (Fig. 18.23). This is not an accident. These cliffs are old wave-cut platforms raised high above sea level long ago by some movement of the earth and are now themselves being consumed by the ceaseless waves.

Topic Estuaries, rias, and fjords

The effects of recent changes of sea level have been dramatic (Fig. 18.20). Estuaries, rias, and fjords are all the result of recent valley drowning.

Rias are drowned valleys in moderate relief. They occur in places such as Chesapeake Bay, USA, Rio de Janeiro, Brazil, and Sydney, Australia. In lowland areas such as eastern Argentina, southeast England, and eastern Canada, the drowned valleys have much more subdued slopes and are called **estuaries**.

Fjords are a distinctive type of drowned valley, partly because they are found only in mountainous regions, but mostly because they were eroded by glaciers rather than rivers. Fjords occur in Scotland (where they are usually called **sea lochs**), New Zealand (where they are called **sounds**), Canada, Norway, and some other coastal mountain areas that have been glaciated. It seems that glaciers were much more effective at eroding down below the low sea level of the Ice Age than were rivers. As a result some fjords, such as the famous Hardanger fjord in Norway, are very deep (up to 1000 m below present water level) and also very long (some are more than 200 km). Further, most fjords are shallower near the coast. In effect they have a lip, called a **threshold,** where the valley glacier ceased eroding as it met the sea.

Figure 18.23 The rising land
These flat-topped cliffs are 70 m above sea level, but they are the remains of an old wave-cut platform.

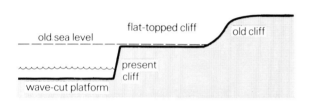

New land from the sea

The growing land

One of the most important jobs in a port is that of marine refuse collector. But the sort of refuse that is collected by port authorities cannot go into the garbage. Every year millions of tonnes of sediment have to be dredged out of our ports so that they will remain open for navigation. It is not a small or insignificant task and it must be continued year in and year out.

Ports are often built at the mouths of rivers because they provide access to inland areas and shelter from storms. But modern ships need access to deep water. Twenty metres is a commonly dredged depth for an international port — often more. If a five-storey apartment block stood in one of these channels, the roof would be below water level.

Of course the river mouth is just where the flow of the river becomes slack and sediment is deposited. In fact, given time a delta would form. So a conflict arises wherever people try to keep a port open. In the long term the river must win, but in the short term continual dredging will keep the channels open. Nowadays there are large dredging boats to deal with the task, but in Europe during the Middle Ages there was no means of preventing silting, and history is full of ports that were abandoned when the river silted up. Famous examples include Brugge, Belgium and the one-time port of Chester in northwest England.

QUESTIONS

Multiple choice

1. Landslides, soil creep, mudflows, avalanches, and slumps are examples of
   (a) mass wasting
   (b) weathering
   (c) erosion
   (d) tectonic processes
2. Leaching is a slow, almost continuous process which may account for at least half of the material removed from a landscape where
   (a) chemical weathering is of little significance
   (b) frost shattering is the dominant weathering process
   (c) slopes are very steep and run-off is particularly rapid
   (d) temperatures are high and moisture is plentiful
3. The portion of a stream's load that is most effective in deepening a stream's bed by corrasion is the
   (a) bed or traction load
   (b) load in solution
   (c) load in suspension
   (d) fastest moving load
4. A meandering path is a feature of natural flowing water, but people sometimes cut meanders off large rivers in order to make the water
   (a) flow more rapidly
   (b) less erosive in its bed
   (c) create ribbon lakes
   (d) deposit its suspended load
5. Complex groundwater caves are common wherever
   (a) coral creates barrier reefs
   (b) erosion due to surface run-off is excessive
   (c) most of the precipitation infiltrates into the soil
   (d) massive beds of limestone have been extensively weathered
6. Alluvial fans and bajadas are rarely significant elements of humid landscapes because they are formed by
   (a) slow, regular deposition of sediments by an exotic stream in an area bounded by faults that has been moved downwards in relation to the surrounding landscape (graben)
   (b) desert winds which cause the evaporation of surface waters
   (c) the outwash of stratified drift from the terminal moraines of glaciers
   (d) occasional flash floods that rush out of arid highlands
7. During each of at least four glacial periods during the past 500 000 years
   (a) ice completely covered all of Canada including the Arctic islands
   (b) sea levels fell by about 100 m while ice covered some northern lands
   (c) the earth's magnetic poles reversed and continental drift ceased
   (d) spreading ice sheets caused the earth's climate to cool by about 6°C
8. Prolonged abrasion and quarrying by alpine glaciers produced U-shaped valleys with
   (a) truncated spurs
   (b) drumlins
   (c) eskers
   (d) alluvial fans
9. Capes and bays have steep and gently sloping shorelines because their particular structures are, respectively, subject to
   (a) tidal waves and longshore drifting
   (b) different levels of tidal change
   (c) swash and backwash
   (d) plunging and spilling waves
10. Estuaries, rias, and fjords are important coastal features caused, in part, by
    (a) erosive work of rivers which run into tropical seas
    (b) flooding of shorelines as ice melted during the past 15 000 years)
    (c) isostasy during the deglaciation of Canada and Scandinavia
    (d) the deposition of sediments in coastal canyons

Short answer

1. Name one example of a waterfall. Explain with labelled diagrams the development of that waterfall.
2. Explain what is meant by the terms river meander, undercut slope, and oxbow lake.
3. Name one example of a delta. Draw and label diagrams to describe the features of a delta and explain how they are formed.
4. What is meant by a swallow hole, a stalactite, and a limestone pavement? Explain how caves and swallow holes form in karst landscapes.
5. Draw and label before and after diagrams that show the formation of an oasis.
6. Define and explain the formation of a cirque, a terminal moraine and a pyramidal peak. Illustrate each feature.
7. Use labelled diagrams to explain the formation of four features of lowland glaciation.
8. Name at least one ria, estuary, and fjord. Draw sketch maps and cross sections of each type of shoreline.
9. Explain why some waves tend to remove sand from a beach while others build wider beaches. What weather conditions are associated with each process?
10. Describe and draw a sketch map of the formation of a bay and headland coastline. Why are headlands often referred to as high energy coasts? What special features are usually found at headlands?

Appendix 1 Latitude and longitude

Where am I, where are you?

Geography is concerned with places on the earth's surface. It tries to show the distribution of natural phenomena and those created by people. But to describe a distribution requires a map, and to construct a map some sort of frame of reference is needed. Communications and navigation would be difficult without a frame of reference; try explaining to somebody the location of Tokyo and its position relative to Vancouver without a map!

There are some natural starting points for our frame of reference. Because the earth rotates about an axis, the points of rotation on the surface can be useful. We call these the **North** and **South geographic poles.** They are nearly in the same places as the **magnetic North and South Poles** — a useful feature as most navigation over long distances is achieved with the aid of a magnetic compass. The **Equator** is another natural reference because it is halfway between the poles and because places north and south of the Equator show many similarities of climate, soils, and vegetation.

An easy start to a frame of reference is achieved by drawing circles on the earth's surface parallel to the Equator. These are **lines of latitude** (parallels) (Fig. A.1). By convention, the Equator is regarded as having a latitude of 0° and the circles parallel to it are designated in degrees north and south until the poles are reached at 90° north and south respectively.

There is no natural starting point for a set of lines in any other direction. By agreement, therefore, a set of circles passing through both poles is spaced evenly across the globe, starting with the circle that passes through Greenwich in London, which is designated 0°. All the other lines (called **lines of longitude** or **meridians**) are measured east and west from this, meeting in the Pacific Ocean. The meridian of 180° east and west is chosen as the International Date Line, as described in Appendix 2.

Combining lines of latitude and longitude produces a grid system with which we can locate all places on the earth's surface. Thus Tokyo is at 35°N, 140°E, and Vancouver is at 49°N, 123°W to the nearest whole degree. We can also say that Tokyo is 1° south of Gibraltar, 1° farther north than Los Angeles, and so on.

We are at grid reference...

Lines of latitude and longitude are useful on maps of the whole world or of continents and large countries, but they are not very convenient for use with maps of small countries such as Britain. So when the British Army Ordnance Corps was given the task of mapping the British Isles in the 18th century, they chose an artificial square grid whose starting point is just off the southwest of Cornwall. The grid is laid over the country and is not related to the lines of longitude in a strict way, although true north and the north of this grid are nearly the same. It is an example of a small-scale system chosen for convenience. Today the grid lines on an Ordnance Survey map are spaced 1 km apart. Most national surveys use the same type of artificial grid.

Figure A.1

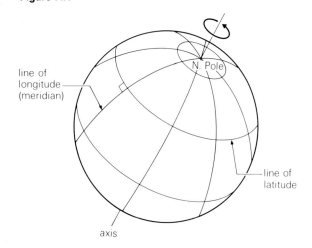

Figure A.2

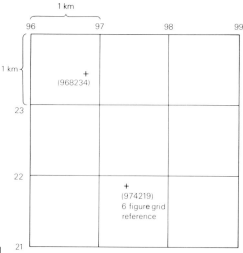

The grid is used to locate places, just as lines of latitude and longitude are. Grid lines running from north to south are called **eastings** and those from east to west are **northings** (that is, east and north of the grid origin). Points on a map are then located by a six-figure reference as shown on Figure A.2. The first three figures are an easting (the vertical line); the last three are a northing (horizontal line). Some people find it easier to remember the order of a reference by saying to themselves "Along the corridor and up the stairs."

Appendix 2   Time

Today is Tuesday...or is it Wednesday?

The **International Date Line,** fixed by international agreement, follows the 180° meridian, except where it deviates to avoid dividing land areas (see inside back cover). The international day begins and ends on this line and sweeps westward with the sun (Fig. A3).

Exactly at midnight (on the date line) it may be Tuesday over the whole globe. Just west of the line it is Tuesday midnight; just east of the line Tuesday is beginning. At every 15°W of the line, it is another hour before midnight. This is because 24 h divide into the 360° of the globe 15 times. Therefore, for example, at 165°W it is 23.00 h Tuesday night and at Greenwich (0°) it is midday; a moment later Wednesday has started to the west of the line; an hour later Wednesday has begun 15°W; and so on...

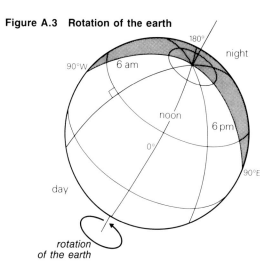

Figure A.3   Rotation of the earth

Appendix 3   Instruments to measure things

Much of the information we have used to build up a picture of the way the earth's surface has been formed and the processes at work in its continued modification comes from direct measurement. Below are some of the pieces of equipment in common use.

Weather

The main elements of the weather that are often measured are air temperature, precipitation, wind speed and direction, duration of sunshine, and pressure. Many of the instruments required for such measurements are best kept in a special box called a **Stevenson's screen** (Fig. A.4). This has slatted sides designed to allow free air movement, but prevent direct sunshine affecting the readings of the instruments. The box also ensures that the instruments are kept a standard 1.5 m off the ground to allow comparison between stations.

Air temperature is measured using a set of thermometers. These are read directly at regular intervals during the day. Usually one thermometer, called a **dry bulb thermometer,** is used for normal air temperature. Another has a wet muslin bag tied over the bulb (a **wet bulb thermometer**). The difference in temperature between them is a measure of the relative humidity of the air. When they read the same the air is saturated. These thermometers are usually mounted vertically within the Stevenson's screen.

In addition a **maximum and minimum** thermometer set is usually provided (Fig. A.5). These record the range of temperatures experienced over a given period. One of these thermometers is usually filled with mercury. A small wire

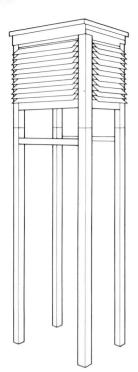

Figure A.4

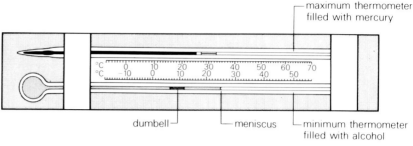

Figure A.5

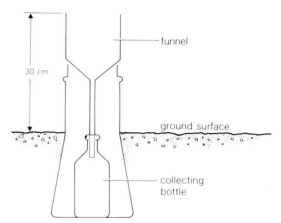

Figure A.6

dumbell in the thermometer tube is pushed forward as the mercury advances and is left behind as temperatures fall, thus recording the maximum value. The minimum temperature is also measured with a wire dumbell, but this time the thermometer is filled with alcohol and the dumbell is dragged along with the liquid as it contracts with lower temperatures. The dumbells can be seen clearly in Figure A.5. Notice that these thermometers are kept horizontally and they need resetting after each reading each day by bringing the dumbell back against the meniscus in each thermometer.

Precipitation is measured with a standard raingauge whose upper rim is placed 30 cm above the ground (Fig. A.6). The contents of the receiving bottle are measured, then emptied, each day. The gauge must not be sited near any sheltering trees or buildings.

Pressure is usually measured using an aneroid barometer which has a recording pen and drum mounted in its case (Fig. A.7). The pen records expansion and contraction of the partially evacuated container X with changes of air pressure.

The duration of continuous sunshine is measured with a recording device comprising a glass sphere which concentrates the sun's rays and burns a trace on a piece of sensitized paper placed in the metal holder (Fig. A.8).

Wind speed is measured with a 3-cup rotating device on a spindle called an anemometer (Fig. A.9). As the wind blows, it turns the cups and the number of revolutions of the spindle is converted into wind speeds in m/s or km/h. A good exposed position away from any buildings or trees is essential.

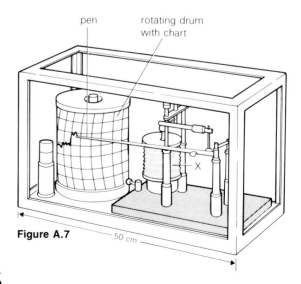

Figure A.7

Water cycle

In addition to weather elements the water cycle requires the measurement of evaporation and streamflow.

Evaporation is measured by a large tank which is daily refilled with water to a marker (Fig. A.10). The amount of water added is equal to the amount lost by evaporation over the previous day.

Streamflow is usually measured with a V-notch weir (Fig. A.11). There is a known relationship between the height of water ponded up behind the V-notch construction (h) and the flow of water. The height is recorded daily or continuously on a drum.

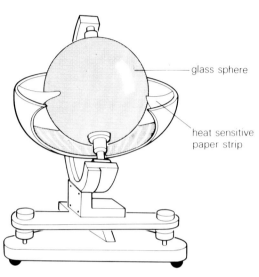

Figure A.8

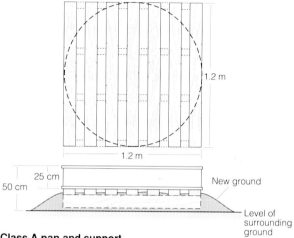

Class A pan and support

Evaporimeters are pans of water from which evaporation can be measured. Because of heat transfers through the sides and bottoms of such pans, evaporation from pans is always higher than from lakes.

The National Weather Service class A pan is 1.2 m in diameter and 25 cm deep. It is supported a few centimetres above the ground on a timber grid. Evaporation from the class A pan must be reduced about 30% to estimate lake evaporation.

Figure A.10

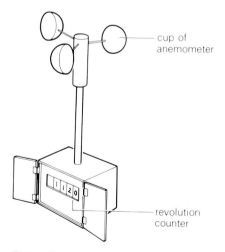

Figure A.9

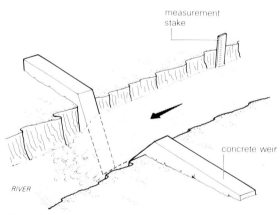

Figure A.11

Glossary

ablation Loss of glacial ice or snow by melting evaporation, or sublimation.

abrasion (corrasion) Bedrock erosion caused by rock particles that are carried by water, waves, wind, or glacial ice.

accelerated soil erosion Removal of weathered materials at a faster rate than normal due to the direct or indirect effects of peoples' behaviours.

acid rain Precipitation with a pH value of less than 5.6. It is most apparent downwind from major urban and industrial areas.

adiabatic heating and cooling Temperature change within a gas due to change in pressure and volume. Compression causes heating, expansion leads to cooling.

advection Horizontal movement of air and energy. Advection fog forms when warm, moist air moves over a much cooler surface.

air mass Large, uniform section of the atmosphere that influences the weather by carrying its distinct characteristics to other areas.

albedo Proportion of solar radiation reflected back into space. The earth's average albedo is about 40%; forests, 5-10%; grasses, 18-30%; and snow and ice, 50-80%.

alluvium Finely weathered rock materials — mostly silt and sand — deposited by water.

andesite Fine-grained igneous rock that forms steep-sided composite volcanoes in mountain chains (e.g. the Andes).

anticline Arch or crest of a fold in rock layers.

anticyclone A region of high atmospheric pressure, also known as a HIGH.

aquifer Porous, permeable rock layer that absorbs and traps ground water.

arch Opening in a mass of steep rock. It is most frequently found in coastal cliffs.

arête Jagged, knife-edged ridge between two expanding cirques or alpine valleys. A low point in an arête is a col; a summit with three or more arêtes forms a pyramidal peak.

artesian well Shaft that is drilled into an aquifer and carries ground water under its own pressure to the surface.

aspect Direction a slope faces. In the Northern Hemisphere, the warmest slopes usually have a southerly aspect.

atoll A ring-like coral reef that surrounds a shallow lagoon. Atolls usually grow on top of submerged, dormant volcanoes.

avalanche Rapid slide of a mass of snow, air, ice, rock, and other debris in alpine areas.

backwash Sea water receding from a beach after the uprush (swash) of a wave.

bajada (bahada) Coalesced alluvial fans at the base of a mountainous region.

barchan (barkhan) Crescent-shaped sand dune with two horns pointing downwind.

barometer Instrument used for measuring atmospheric pressure.

barrage Massive structure that holds back water (e.g. blocks storm surges along low-lying coastlines or retains fresh water streams used for irrigation).

barrier reef Extensive coral reef parallel to a coastline. Australia's Great Barrier Reef is the world's largest barrier reef.

basalt Heavy, dark, fine-grained volcanic rock that shrinks as it cools, cracking into vertical columns. Shield volcanoes are composed largely of basalt (e.g. Hawaii, the big island).

batholith Large, dome-shaped intrusion of igneous rock, usually granite.

biosphere The part of planet earth that sustains life.

bore Small, predictable tidal wave that moves rapidly up a shallow bay or river. Famous bores run up: the Petitcodiac River, New Brunswick; Severn River, Wales; Sepik River, Papua New Guinea.

boreal forest Northern coniferous forest dominated by spruce, fir, and pine trees.

braided stream River channel split into many sub-channels by loose alluvial deposits.

breakers Waves that break as foam in shallow water. The shallow water slows their progress.

caldera An extensive depression caused by erosion, explosion, or implosion of a volcano.

canyon A gorge that is large, steep-walled, and water-worn. The largest and best known is the Grand Canyon of the Colorado, Arizona, USA.

capillary water Soil water that bonds by surface tension to soil particles. It moves in any direction from areas of surplus to areas of deficit.

cartography Science and art of map making.

chernozem A grassland soil (sometimes called black earth) that is very dark, fertile, and alkaline. It crumbles easily and is ideal for grain growing.

chinook A warm wind that undergoes adiabatic heating. This westerly wind is dry and blows down the eastern side of the Rockies.

cirque (corrie, cwm) Deep, steep-sided hollow formed by glacial erosion at the head of an alpine valley.

cirrus Thin, wispy, ice-crystal clouds that are high (6-12 km) above the earth's surface.

clay Finest particles of rock (>0.002 mm in diameter). Clay tends to form porous, but impermeable B layers in mature soils.

climate Long-term average weather of a region.

climax vegetation Best vegetative community for the climate, soil, topography, and ecological condition of an area.

cloud Floating mass of visible water droplets. Condensation nuclei and water vapour are required for cloud formation.

cold front Steep-sloping, leading edge of a cooler air mass as it overtakes a warmer, less dense air mass.

condensation Process which transforms vapour into a liquid, thereby releasing the latent heat of evaporation.

conduction Heat transfer from molecule to molecule. It is caused by internal molecular movement.

coniferous forest (boreal forest) Mainly cone-bearing trees that are most widespread north of 50°N (Canada, Europe, and Asia).

continent Large land mass which includes submarine continental shelves. The shelves rise abruptly from oceanic plains. Australia is a continent.

convection Heat transfer caused by circulation in a gas or liquid. **Convectional precipitation** might occur when heated air rises significantly.

crag and tail Glacial remnant of resistant bedrock (crag) that has a lee side tail of gravel or softer rock. One of the best examples is Edinburgh Castle (crag) and the Royal Mile (tail) in Scotland.

crevasse An almost vertical stress crack in a glacier.

crust The outer shell of the earth which is relatively thin (8-64 km), brittle, and has a low density. It is composed of SIAL (continental) and SIMA (oceanic) rocks.

cumulonimbus Large, dense storm cloud with extreme vertical development (8-20 km). Thunder, lightning, heavy rain, and/or hail are often associated with these clouds.

cumulus Flat-bottomed, dense, globular clouds that often signal fair weather.

cyclone Area of low atmospheric pressure or a LOW. Winds blow into cyclones.

deciduous forest Usually broad-leaved trees that shed leaves in the autumn or at times of greatest environmental stress.

deflation Wind removal of loose, dry material from the land surface.

degradation Reduction of land surface by erosion towards the ultimate base level (sea level).

delta Almost flat area of alluvium at a stream's mouth. It exists where a number of distributaries flow into a body of still water.

dendritic drainage An irregular, tree-like branching pattern of tributaries.

depression (i) a basin or hollow with internal drainage (e.g. Qattara, Egypt); (ii) an atmospheric LOW pressure area; a cyclone.

desert An area of severe water deficit.

desertification Expansion of desert areas due to processes such as climatic change and cultural effects including global warming, deforestation, over-grazing, and wildfires.

dew Deposit of water droplets on cold surfaces after the air temperature drops below dew point.

dew point Temperature at which cooled air achieves saturation.

dike More or less vertical intrusion of igneous rock which cuts across the bedding planes of other rocks. (*See* **sill**)

diurnal range Difference between the highest and lowest values within a 24-hour period.

drift (i) All materials deposited by a glacier including the outwash materials; (ii) sand mass with no specific shape.

drumlin Stream-lined, longitudinal hill of moraine with tail pointing in the direction of ice advance.

dune Sand mound deposited and shaped by the wind to a characteristic form.

ecology Study of the interactions of organisms and environments.

environment Surroundings, particularly the factors that affect the growth and development of organisms.

epicentre Point on the earth's surface directly above the centre (also called seismic focus or **hypocentre**) of an earthquake.

equinox Either of the two times (21 March and 22 September) when the sun is directly over the Equator. It causes day and night to be of equal length.

erg A vast expanse of sand and dunes.

erosion Transportation of weathered materials by running water, waves, wind, or ice.

erratic Boulder transported and deposited by a glacier.

esker Narrow, winding gravel ridge believed to have been deposited by a sub-glacial stream.

estuary Extensive, V-shaped river mouth where fresh and sea water mix (e.g. St. Lawrence, Canada; Thames, England; Rio de la Plata, Argentina/Uruguay).

evapotranspiration Total water loss from soil and water surfaces (evaporation) and plants (transpiration) to the atmosphere.

extrusive rock Igneous rock solidified from magma which has poured onto the earth's surface.

fallow A farming technique that leaves the land unseeded to assist with fertility recovery.

fault Fracture in the earth's crust along which rocks have moved relative to each other. Normal faults are caused by tension which leads to the formation of escarpments and rift valleys.

fauna All animal life available to a region.

fertilizer Natural or artificial substances used to increase the productivity of the soil.

firn (névé) Compacted granular snow that may lead to formation of glaciers.

fissure Long break in the earth's crust from which volcanic materials occasionally extrude.

fjord Deep glacial trough flooded by sea water (e.g. Indian Arm, Howe Sound, Rivers Inlet, BC).

flood plain Lowland alluvial area subject to flooding by its river. Common features include: levées; oxbow lakes; meanders; backswamps.

flora All plant life available to a region.

focus (hypocentre) Point within the earth's crust where an earthquake occurs. *See* **epicentre**

fog Ground-level mass of suspended water droplets.

fold mountains Created by the compression and folding of massive rock strata (e.g. Himalayas, Andes, Alps, Rockies). *See* **geosyncline**

front Sloping atmospheric boundary between different air masses.

frost Deposit of minute ice crystals caused by water condensation at or below 0°C. Frost shattering by the expansive power of cooling water is a powerful weathering agent.

geo Borrowed from Greek, it means of the earth.

geography Study of the earth as the home of people.

geosyncline Vast, roughly longitudinal depression where sediments accumulate. Compressions of geosynclines cause fold mountains.

geothermal Heat from within the earth. Geysers indicate geothermal activity.

glacier Large ice mass which flows from a firn field. Glaciers may be either sheets or rivers of ice.

gorge Narrow, steep-sided river valley (e.g. Niagara Gorge, Ontario).

graben Long, narrow rift valley between at least two normal faults.

granite Common, coarse-grained plutonic rock composed of quartz, feldspar, and mica.

gravel Mixture of rock fragments broken to 2-64 mm in diameter.

Great Circle Any circle on the earth's surface which has the earth's centre as its centre. An arc of a Great Circle is the shortest distance between two points on the surface of the earth.

greenhouse effect Warming of earth's atmosphere that occurs when outgoing long-wave radiation is absorbed by water vapour and carbon dioxide. It began billions of years ago. Recently, the burning of fossil fuels has added 30% more carbon dioxide to the atmosphere, adding to the greenhouse effect.

gully Narrow, steep-sided channel that is worn by water.

hamada Rock desert of bedrock pavement. (*See* **reg**)

hanging valley Tributary U-shaped valley entering a larger glacial trough well above the floor of the larger valley.

headland A cape in a lake, sea, or ocean that ends with a steep cliff.

HIGH (anticyclone) An area of high atmospheric pressure. Winds blow out of HIGHS.

horn Pyramidal peak that is formed in glaciated landscapes when three or more arêtes meet at a summit. Horns are surrounded by cirques.

horst Longitudinal upthrust mountain block that is bounded by at least two normal faults.

humus Dark, partly decayed organic matter which is vital to the binding and fertility of topsoil.

hurricane Severe cyclonic storm (wind speeds > 160 km/h) in the Americas (known as typhoons in Southeast Asia and cyclones in Australia).

hydrologic cycle Continuous circulation of all water in the atmosphere, water bodies, and land. Precipitation = Evaporation + Transpiration + Run-off + Infiltration

igneous rock Any rock formed directly from a molten state.

intrusive rock Igneous rock that solidified deep within the earth's crust. It usually forms large masses of plutonic rock such as granite.

inversion An anomaly that occurs when atmospheric temperatures rise with increasing altitude.

isobar Line on a map that joins points that have the same atmospheric pressure.

isostasy Condition of equilibrium between sections of the earth's crust. Isostatic readjustments occur when extensive land areas (e.g. Canada) are unweighted due to deglaciation and/or erosion or weighted (e.g. Netherlands) due to sedimentation and/or increasing ocean depth.

isotherm Line on a map that joins points with the same atmospheric temperature.

jet stream High velocity (120-160 km/h) winds in the upper atmosphere (6-12 km), especially in the mid-latitudes.

katabatic wind Cold downslope wind that occurs usually at night. It is caused by the settling of colder, more dense upland air.

kilopascal Measure of air pressure, equal to 10 millibars. Normal sea level air pressure is 101.32 kPa.

lagoon Shallow body of salt water separated from the sea by a narrow reef, spit, or bar.

land breeze Nighttime convectional breeze that blows gently from land to sea.

landslide Massive, usually rapid downslope movement of weathered earth materials.

latitude Angular distance north or south of the Equator measured in degrees to as far as 90° N/S.

lava Molten material extruded from a volcanic vent. Lava solidifies to form extrusive igneous rock.

leaching Removal of soluble substances from rock or topsoil by infiltrating water.

limestone Rock consisting of at least 50% calcium carbonate.

loam Arable topsoil consisting of about 20% clay, 40% silt, and 40% sand.

lode Vein or several parallel veins of mineral within a rock mass.

loess Wind-deposited silts that are found usually downwind from arid or glacial areas.

longitude Angular distance measured in degrees east or west of the Prime Meridian to as far as 180°E/W.

longshore drift Zig-zag movement of weathered materials along a beach. It is caused by the swash and backwash of oblique waves.

LOW An area of low atmospheric pressure; a cyclone. Winds blow into lows.

magma Molten rock materials beneath the earth's surface.

mantle Massive, concentric layer of the earth, about 2900 km thick, consisting of relatively dense, hot rock between the crust and the core.

mass wasting (mass movement) Downslope movement of weathered rock materials largely as a result of the earth's gravitation.

metamorphic rock Rocks reformed from any other rocks by heat, pressure, and/or fluids. Extreme heat causes the rock to melt, creating igneous rocks.

meteorite Rocky remnants of meteors that reach the earth's surface. Meteorites are usually rich in iron and nickel or silica.

millibar Common unit of measurement of atmospheric pressure. Normal sea level air is 1 013.2 mbs.

mineral (i) In general usage, a substance obtained by mining (e.g. gold, coal, oil); (ii) scientifically, a natural inorganic, homogeneous substance with definite characteristics of hardness, colour, and form.

monoculture Cultivation of a single type of crop on a farm, plantation, or in an agricultural area.

monsoon Seasonal wind, usually within the tropics, blowing roughly opposite to either the northeast trades or southeast trades.

moraine Unsorted rock debris transported by ice and deposited by the melting of glaciers.

mudflow Mass wasting of saturated soil.

mulch Layer of organic debris covering and protecting soil.

névé (firn) Snowfield or area of ice accumulation that leads to glacier formation.

nimbostratus Extensive layer of low-level rain cloud.

normal fault Generally vertical displacement of sections of the earth's crust along a deep break due to tension in the crust. The tension acts at right angles to the line of the fault.

occluded front Very active weather system. It occurs when a cold front overtakes a warm front and forces the intervening warm air aloft.

ocean current Large-scale horizontal movement of ocean water due to winds, salinity, and the Coriolis effect. Ocean currents move clockwise in the Northern Hemisphere.

orographic (relief) precipitation Precipitation following rapid condensation in moist air that undergoes adiabatic cooling when forced to rise over mountains.

outwash Glacial drift washed by melt water beyond a terminal or recessional moraine.

overfold An extremely compressed, assymmetrical anticline.

oxbow lake Crescent-shaped lake formed when a meander is cut off from the course of a slow-flowing river.

pack ice Extensive, relatively thin ice mass that covers part of a sea or ocean.

peat Deep, partially decomposed mass of organic materials usually accumulated in swamps or bogs. Peat is frequently cut, dried, and burned as a low-grade fuel.

pebbles Weathered, rounded stones that are 2-64 mm in diameter.

permafrost Permanently frozen soil typical of tundra and high alpine regions.

plantation Extensive area devoted to cash crops and monoculture. Crops include cotton, tea, coffee, rubber, and sugar cane.

plate tectonics Concept that replaced continental drift theory. It states that the earth's crust consists of six major, separate, and dynamic plates.

playa Dry lake bed in an arid area, usually white with salts.

plucking (quarrying) Glacial erosion process whereby protruding or fractured rock surfaces are frozen to the glacial ice and broken free from bedrock.

pluton Extensive mass of igneous rock formed by slow cooling deep within the earth's crust (e.g. Sierra Nevadas, California).

podzolization Soil formed by a leaching process in humid mid to high latitudes and altitudes.

pyramidal peak (horn) Sharp rocky mountain peak caused by the active headward erosion of three or more cirques.

pyroclastic materials Cinders, ash, and rock fragments thrown into the air by vigorous volcanic eruptions.

radiation Emission of waves or particles through space (e.g. solar radiation).

radiation fog Thin fog that may occur when air is cooled by contact with a cold earth surface. It happens especially at night when energy is being radiated to space.

rain shadow Phenomenon experienced on the dry lee side of a mountain where descending air undergoes adiabatic heating.

rainsplash Breakup and movement of loose soil particles by the impact of raindrops.

rapids Fast, turbulent flow of a stream over a steep, uneven bed.

reg An extensive desert lowland covered by pebbles and gravel.

region A distinctive area of the earth's surface defined by its particular characteristics.

relief Difference in height between high and low points in a region's topography. **Relief precipitation** occurs where moist air rises over mountains.

resources Anything of any use to anybody. **Renewable resources** are regenerated (soil, water, trees). **Non-renewable resources** cannot be replaced (oil, coal). Other resource categories include scarce, abundant, human, and natural.

ria Complex sea inlet caused by oceanic flooding of a river valley.

ribbon (finger) lake Glacial-trough lake that is long, deep, and narrow, and usually dammed by a moraine (e.g. Okanagan Lake, Harrison Lake).

ridge (i) Elongated, narrow upland area; (ii) in meteorology, a narrow area of HIGH air pressure.

rift valley Long, narrow, down-faulted trench between normal faults and upthrust horsts. Famous rifts include: Death Valley, California; Rocky Mountain Trench, BC; and the Dead Sea, Israel/Jordan.

rills Small, water-worn channels that tend to grow and form gullies.

roche moutonnée Rocky outcrop shaped by a glacier. It is abraded and polished on the upstream side and has plucked cliffs on the downstream side.

rock fall Almost vertical drop of weathered rock from a cliff. Rock falls form screes.

salinization Accumulation of soluble salts in topsoils of arid areas due to evaporation of both surface and capillary water.

sand Rock particles from 0.02-2.0 mm in diameter.

saturation When air can hold no additional water vapour. At saturation, relative humidity is 100%.

scree (talus) Steep slope of coarse, sharply fractured rocks at the base of a mountain slope.

sea breeze Afternoon convectional breeze blowing briskly from sea to land.

sea floor spreading Movement of new oceanic crust away from oceanic ridges such as the Mid-Atlantic Ridge.

sedimentary rock Rock formed by the accumulation, compaction, and/or cementation of weathered rock particles (e.g. clays, silts, sands, gravels, and evaporites).

seif dune Long, narrow, steep sand dune that is parallel to the prevailing wind direction.

shield volcano Volcanic complex created by numerous relatively quiet eruptions of fluid basaltic lavas (e.g. Hawaii).

sill More or less horizontal intrusion of solidified magma between natural layers or bedding planes of rocks. (*See* **dike**)

silt Fine rock particles from 0.002-0.02 mm in diameter.

slip-off slope Gentle slope of unconsolidated sediments on the inner, convex side of a river meander.

smog Fog contaminated by smoke, dust, carbon monoxide, and other fumes. Most common and severe in urban and industrial areas.

soil The part of the earth's surface that sustains the growth of plants. Soil consists of weathered rock fragments, decaying organic matter, air, and water. Mature soils develop clear layers or horizons.

soil creep Slow downslope movement of soil due mostly to gravity.

soil profile Vertical cross section that exposes soil layers (horizons).

solstice Either of the two times (22 June and 22 December) when the sun is at the greatest distance from the Equator.

solution Process of chemical weathering of solutes from rock.

stack Rocky vertical island, often a pillar, that is a remnant of sea cliffs.

stalactite Fine, conical column of calcium carbonate that hangs vertically from the ceiling of a limestone cave.

stalagmite Rough, broad, cone-shaped accumulation of calcium carbonate that grows upward from the floor of a limestone cave.

storm wave Large, usually destructive wind-driven ocean wave.

storm surge A general rise in sea level (as much as 5 m) during a cyclone. It is due to onshore wind and an intense LOW air pressure near the cyclone's centre.

strato-volcano (composite cone) Volcanic cone formed from alternating pyroclastic materials and lava layers erupted from a central vent. Most well-known volcanoes are formed in this manner.

stratus Low, relatively uniform cloud layer. Poor illumination causes it to frequently appear grey.

stream load Total amount of material carried by a stream. It includes solutes, suspended, and bed loads.

striations Grooves and scratches on bedrock surfaces caused by rock abrasion under glacial ice.

subduction An essential process of plate tectonics that occurs when the thin oceanic crust (SIMA) moves under a thicker continental plate (SIAL).

swallow hole (doline) Large funnel-shaped depression in a limestone landscape.

swash Frothy, broken water that surges up a beach after a wave breaks.

swell One of a group of long duration, regular waves.

syncline Downfold in the earth's crust caused by compression.

tarn Small, round lake on the floor of a cirque.

tear (horizontal or slip-strike) fault Generally horizontal displacement of rocks found along a break in the earth's crust (e.g. San Andreas Fault, California).

terrace Gently sloping shelf of sedimentary materials bounded above and below by steep slopes.

terracette Small shelf of sediment or soil materials — 10-50 cm high — on a hillside where slumping or soil creep has occurred.

terrestrial An adjective that refers to earth or land.

tide Periodic, predictable rise and fall of sea level due mainly to gravitational effects of moon, sun, and the earth and its rotation.

till Compact, but unsorted and unstratified deposits of rock debris. It ranges from clay to boulders.

tombolo Bar, usually of sand, on the lee side of an island. A tombolo ties the island to the mainland.

topography Description or map portrayal of a place's physical features.

tornado Small, violent whirlwind with extremely low air pressure at its centre, and wind speeds > 320 km/h.

trade winds Steady winds blowing from sub-tropical HIGHS to inter-tropical LOWS. They are northeasterly in the Northern Hemisphere and southeasterly in the Southern Hemisphere.

tributary Stream that contributes water to a larger stream.

tropical rain forest (selva) Very complex, luxuriant, forest of the tropics. Trees are broad-leaved evergreens with many vines, ferns, and orchids.

truncated spur Ridge cut off by an alpine glacier.

tsunami (seismic wave) Large, long-period ocean wave caused by submarine earthquake or volcanic activity. It is imperceptible at sea, but catastrophic on shorelines, especially in bays or inlets.

tundra Treeless arctic or high alpine area where no vertical trees grow.

turbulence (air) Irregular, often violent, airflow caused by differential heating of the atmosphere or strong frontal activity.

undercutting Erosion of the base of (i) an outer bank of a river meander; (ii) a sea cliff; (iii) a desert rock pedestal (wind).

valley train Outwash deposits in a glacial trough. It resembles a long, narrow alluvial fan.

veins Crystallized metal ore found in a rock mass fissure.

wadi (canyon, arroyo) Steep, water-worn rocky ravine in a desert.

warm front Leading edge of warmer, less dense air mass overtaking a cooler, more dense air mass.

waves (ocean, sea, or lake) Wind-driven oscillations of a water surface. On steep shorelines waves *plunge*, causing erosion, but on gently sloping shores they *spill* slowly, dragging sand onto the shore.

wave-cut platform Gently sloping ledge at the base of a cliff which has been undercut by wave action.

weather Momentary atmospheric condition at a specific place.

weathering Physical disintegration and/or chemical decomposition of rocks and minerals.

Sources and Credits

A wealth of technical data were provided by Alan Nourse, Environment Canada; Jack Cann, Quinsam Coal; John Harper, Memorial University, Centre for Earth Resources Research; David Jarman, Adelaide; Silvana Harwood, Denison Mines Ltd.; John Toye, Manitoba Water Resources Branch; Marcia Knapp, Sullivan Mine; and Peter Adams, Trent University. Excellent photographs were provided by Ian Snow, Harvie Walker, David Williams among others. Maps are based on information from acknowledged sources.

Roger Jones = RJ; Brian Knapp = BJK; Victor C. Last = VCL; H.L. Walker = HLW; David Worrall = DSW

A John Mack; B RJ; 1.0 USGS; 1.4 RJ; 1.8 (both) Gary Rosenquist/Earth Images; 1.9 (left & right) Stephen Porter; 1.10 Stephen Porter; 1.16 BJK; 1.17 DSW; 1.19 Washington State Tourism Division; 1.20 HLW; 1.21 Carnegie Institute of Washington Pub. No. 540, 1942, p. 104, Figure 29; 1.22 BC Ministry of Energy, Mines, & Petroleum Resources; 1.24 (left) Timothy Goulter (right) RJ; 1.26 (left & right) BJK; 1.27 Sullivan Mine, Kimberley, BC; 2.1 BJK; 2.5 Bruce Coleman Ltd.; 2.2 *The Western Canadian Sedimentary Basin,* edited by Glynn N. Wright, Canadian Society of Petroleum Engineers; 2.9 National Atlas of Canada; 2.10b Saskatchewan Department of Mineral Resources; 2.11a&b *Canadian Mining Journal,* Nov. '84, pp. 30-31, "All dressed up and looking for a place to go," *Mining Review,* July/Aug. '84, & J. Cann, Quinsam Coal; 3.2 BJK; 3.6 VCL; 3.9 RJ; 3.10 RJ; 3.14a Quintette Coal, Tumbler Ridge, BD; 3.14b&c *Canadian Mining Journal,* Nov. '83, pp. 39-51, "A Geological Review of Denison's Coal Properties, Sagi & Krishnan; 4.2 Based on *Physical Geology,* Arthur N. Strahler, 1981; 4.7 American Geophysical Union; 4.8 National Atlas of Canada, p. 26; 4.9 *Scientific American,* May '60, Vol. 202, No. 5, pp. 70-79, "The Changing Level of the Sea," Fainbridge; 5.0 NASA; 5.1 Ann Ronan Picture Library; 5.4 German Wine Institute; 5.11 E. Yablonski; 5.12 HLW, Environment Canada; 5.19 University of Dundee; 5.25 & 5.26 D.J. Jarman; p. 70 *Victoria Times-Colonist,* 14 October 1984; 6.5 METEOSTAT photographs supplied by the European Space Agency; 7.4 (left) VCL; 7.4 (right) Brian Milne/First Light; 8.1 Manitoba Water Resources Branch; 8.1 (inset) Manitoba Archives; 8.2 Manitoba Department of Natural Resources; 8.4 BJK; 8.9 VCL; 8.11 BC Hydro; 8.18 Four by Five/Shostal Associates, New York City; p. 106 VCL: Doug Fugger, Lester B. Pearson College of the Pacific, Victoria, Atlantic Salmon Federation, Four by Five, VCL; 9.0 (left) VCL; 9.0 (right) Aubrey Diem/First Light; 9.1 Ian Snow; 9.2 BJK; 9.6 DSW; 9.7 BJK; 9.8 DSW; 9.9 DSW; 10.2 John Stout; 10.4 BJK; 10.8 Raymond Jessop; 10.9 BJK; 10.11 BJK; 11.4 VCL: 11.7 BJK; 11.8 DSW: 11.10 Rainforest Action Network; 11.12 Raymond Jessop; 11.13 John Mack; 11.14 John Burn; 11.15 BJK; 11.16 DSW; 11.17 VCL; 11.18 HLW; 11.19 DSW; 11.22 HLW; 12.1 Four by Five; 12.2a John Mack; 12.2b&c Ian Fenwick; 12.2d Raymond Jessop; 12.5 BJK; 12.6 Ian Fenwick; 12.7 USDA; 12.11 Raymond Jessop; 12.13 Four by Five/Shostal Associates, New York City; 12.15 Four by Five/Shostal Associates, New York City; 13.0 Denys Brusden; 13.1 VCL; 13.3 BJK; 13.4 BJK; 13.5 A. Matthews; 13.8 VCL; 14.1 VCL; 14.6 BJK; 14.7 BJK; 14.8a Ian Snow; 14.8b Four by Five; 14.10 Image Bank; 14.11 BJK; 14.13 (left) BJK; 14.13 (right) Capilano Canyon Suspension Bridge; 14.15 BJK; 14.18 BJK; 14.22d Wayne Duchart; 15.1 BJK; 15.2 VCL; 15.4 BJK; 15.5 VCL; 15.5 Ian Snow; 16.1 Bernard Smith; 16.3 BJK; 16.4 BJK; 16.5 BJK; 16.6 BJK; 16.7 Mark Burnham/First Light; 16.9 Harvey Walkland; 17.1 Jessie Parker/First Light; 17.2 VCL; 17.5 A. Tereposky; 17.6 BJK; 17.10 BJK; 17.12 EME; 17.13 BJK; 17.14 DSW; 17.21 VCL; 18.2 Brian Milne/First Light; 18.7 Memorial University, Centre for Earth Resources Research; 18.8 I. Snow; 18.11 DSW; 18.12 Memorial University, Centre for Earth Resources Research; 18.14a DSW; 18.14b HLW; 18.16 BJK; 18.20 Brian Milne/First Light; 18.23 BJK.

Index

Boldfaced words are defined in the glossary.
Boldfaced page numbers refer to illustrations or photos.

abrasion
  ice, 207, 213
  water, 174
  waves, 225
  wind, 200
acacia tree, **130**
accelerated soil erosion, 143
acidity, 119
adaptation by vegetation, 125-126
advection fog, 65
air pressure, 57, 63
  temperature, 50, 60
Alberta, 30, 80
alluvial fan, **196**, 198
alluvium, 116, 181, 187
Alps, 42, 43, 45, **137**
  mountain building in, **40**, 42
altitude, **137**, 205
Andes, South America, 11, 41-42, 45
andesite, 11, 16, 41
anemometer, 236-**237**
anticline, 22, 36-37
anticyclone, 61, 63, **68**
Appalachians, 2-3, 30, 36, 46
aqueduct, **97**
aquifer, **93**, 99
arch, 221-**222**, 225
arête, **210**-211, 213
artesian well, 93
ash, 11-13, 16, 20
aspect, 52
Assiniboine River, **89-90**, 92
Aswan High Dam, **104**
atmosphere, 50-70
atoll, 26, **27**-28
avalanche, 165, 211

backwash, 223, **227**
bajada, 197
barchans (sand dunes), **201**
barometer, 63, **236**
barrage, 85, 232
barrier reef, **27**-28
basalt, 9, 16, 46
batholith, 11, **15**-16, 24, 35, 44, 46, 112, 131
bay, 221-**222**
beach, 224-226
bedding plane, **112**
bloom, of plankton, **87**
braided stream, **180**-181
breakers, 219
British Columbia climate, **72-73**
brown earth soil, **117**, 120
building materials, 29
Bumpass Hell, USA, **21**
bush fallow system, 140-142
buttress tree, **127**

cactus, **125**
caldera, 17-**18**
Canadian Shield, 24, 35, 46
canyon, **178**, 198-199
carbon dioxide, 51, 113
Cascade Mountains, 16, 17, **19**, 21, 134
Cathedral Lakes, BC, 134-**135**
cave
  ice, **207**
  limestone, **192-193**
  sea cliff, **221**, 223, 226
cement, 29, 32
chalk, 93, 194
Chamonix, France, **203**
chemical weathering, 108, 111-114, 116, 164, 166, 225
chernozem soil, **118**, 120, 218
cinder cone, **9**, 18
cinders, 16
circulation
  of atmosphere, 54, 56
  of ocean, 82
cirque, **208-209**, **210-211**, 213
cirrus, **58-59**, **61**, 68
citrus farming, 149
clay, 10, 25, 113-114, 116, 119
cliff, **220-221**, 223
climate, 71
  cold, 78, 79
  continental, 73
  cool temperate, 71, 73, 79
  desert, 50, 60, 76, **78-79**
  eastern margin, 79
  equatorial, 76, 79
  hot, 71, 76, 79
  maritime, 73
  Mediterranean, 79-80
  monsoon, 74, 76, 78-79
  tropical, 79
  warm temperate, 79
climatic change, 80
  regions, 71
climax vegetation, 127, 129
cloud, 57-59, 61. See also specific types
coal
  fields, 24, 30
  formation, 30, 46
  mining, 32, 39
coastal breeze, **55**
coffee, 150-151
cold front, 68
cold ocean current, **82**-83, **84**, 88
colonization by plants, 125, 128
Colorado River, USA, 23, 97-**98**
Columbia River, 13, 97
concrete, 29
condensation, 53
conduction, heat, 53-**54**
coniferous forest, 32, 131-**132**, 138, 142

conservation
  soil, 146-**147**
  water, 90, **150**
constructive waves, **226**
continental drift, 40, **45**, 133
contour ploughing, **147**
convection
  in atmosphere, **53-55**, 67
  in earth, 7-**8**, 40
  in oceans, 82-83
  causing precipitation, 67
coral reef, **26-28**
core of earth, 8
Coriolis effect, 69
corrie. See cirque
crag and tail, **208**
Crater Lake, USA, **18**
crevasse, **208-209**, 211
crust of earth, 8, 16, 40-41
cuesta, **171**
cumulonimbus, **58-59**, 61, **66-67**
cumulus, **58-59**, 61, 66
curtain stalactites, **192-193**
Cyclone Tracy, 69

Dead Sea, 31, 34
Death Valley, **198**
deciduous forest, 131
decomposers, 115
deflation armour, 200-**201**
delta, **188-190**
dendritic drainage, **171**
depression, in atmosphere, 61, 63-**64**, **68**, 71
desert, 111, 195
  cold, 79
  farming, 149
  hot, 79
  vegetation, **125**-127, **132-133**, 197
desertification, **78**, 153, 155, 202
destructive waves, **226**
dew, 57
dikes
  for flood control, **90**
  of igneous rocks, 9-11, **15**-16
  on river banks, **184**, 187
dip slope, 171
disaster
  drought, 80, 89-90, 96
  earthquake, 33
  flood, **89-90**, 172
  landslide, **166**
  marine erosion, 223
  soil erosion, 154
  volcanoes, 5, **12-19**
doldrums, 56
drainage basin, **90-91**, 93, 100, 170-**171**
drought, 75, 80, 89-90, 96, 98
  effects on vegetation, 125-126, **133**
  effects on water cycle, 89-90, 96
drowned valleys, 230

drumlins, 214-215, 217
dry-bulb thermometer, 235
Dust Bowl, **143**-144, 146

earth
  origin of, 6
  structure, **8**
earthquakes, 10-11, **33**-34, 37, 40-41, 44, 46
eastings, **234**-235
ecosystem, **107,** 121, **122**-123, 133, 136, 138-149
effluent, 101
energy, solar, 51
equinox, 72
erg, 199-200
erosion
  by ice, 207
  by river, 175
  by sea, 220
  by wind, 200
  of soil, 143
  prevention, **147**-148
erratic, 206, 214
esker, 213, **215-216,** 217
estuary, 230, 232
evaporation measurement, **237**
evaporites, 30-**31**
evapotranspiration, 98-99
evolution, 121-**123**
exfoliation, **110-111,** 113
extrusive rock, 16

fan. *See* alluvial fan
farming
  high technology, 149
  primitive methods, 102, 138, 151
  the past, 139
faults, 20, 21, **33**-34, 36-37
fertility of soil, 142
fertilizers, 31-32, **100**-101, 142-**143**
fire-resistant plants, 133
firn, 206
fishing grounds, **82,** 86-**88**
fissures, 9, 14, 22, 44
fjord, 230, 232
flash flood, **195,** 198-199
flood plain, 94, 181, 184, 186
flooding, 13, **89-90,** 92-94, 98, 187
Floodway, Winnipeg, **90,** 92
flora and fauna, 134
flowstone, 192-**193**
fog, 65
fold mountains, 35, 42, 44
folds, 35-37
food chain, **123**
fossils, 24, 86
freeze-thaw. *See* frost shatter
fringing reef, 26-27
front
  in atmosphere, **68**
  in oceans, 83
frontal cloud, **62**
  precipitation, 62
frost, 150
frost shatter, 108-109, 113, 210
frost-shattered peak, 210

Ganges delta, 190
gas, **24, 30,** 32, 46
geosyncline, 31, 46
geothermal power, 22
geyser, 20-21
glacial
  erosion, 207
  landforms, **203**-218
  valleys, 203, **210,** 212
global winds, **56**
gneiss, 10, 44
gorge, 24, 178-**179**
granite, 10-11, 15, **22,** 24, 35, 37, 112, **175**
Grand Banks, Newfoundland, 31, 84, 87-**88**
Grand Canyon, USA, 23-24
Grand Coulee, USA, 97
grassland, **130**
gravel, 32, 114
Great Barrier Reef, Australia, **27**-28, 121
Great Glen Fault, Scotland, **34**
Great Lakes, 36, **185**
Great Plains, 72, **143**-144, 146, 148
Great Salt Lake, Utah, USA, 31
Great Whin Sill, England, **15**
greenhouse effect, 51
Greenwich meridian, **242**
grid reference, **234**
ground moraine, 207-**208,** 212, 217
ground water, 93, 99
groynes, **229**
Gulf Stream, Atlantic Ocean, 54, 82-84, **88**
gully, 145, 154, **163,** 167, **169,** 171
gypsum, 29, 31

habitat, 122
hamada, 195, 201
hanging valley, 177, **210**-211, 213
hard pan, 120
headland, **221,** 225
heat
  conduction, 53-**54**
  power of sun, 51
Heimaey, Iceland. *See* Iceland
Helgafell, Iceland. *See* Iceland
high pressure regions, 55, 61, 63, 68
Himalayas, 42-**43,** 45-46, 77
horse latitudes, **56**
horst, 37
hot springs, 19-22
humus, 114, 119, 146, 148
hurricane (cyclone), 54, **69,** 78-79
hydro-electric power, **183, 208,** 211
Ice
  bergs, 83, **84,** 86
  cave, **207**
  fall, **209**
  sheet, **204,** 214, 218
Ice Age, 204
Iceland, 9-11, **14,** 21-22, **43,** 45
  Heimaey, 8-9, **14**
  Helgafell, 9-10, **14,** 19, 44
igneous rock, 10, 16, 23-24, 29, 112
India, 42, **45,** 77

monsoon, 77
infiltration, 91
insolation, 51
interglacial period, 204
international date line, **235**
intrusive rock, 16
iron ore, 30-31
irrigation, 101, **102-103,** 150
island arc, 42
isobar, 63
isostasy, 47
isotherm, 72-73

jet stream, 54, 56, 70, 77-78
joints, 37, **112**
jungle, 129

Karst Plateau, Yugoslavia, 27

lagoon, 26, 27, 28, 31
Lake Eyre, Australia, 31
land breeze, 55
landslide, 157, 159, 160, **161,** 163, 167, **169**
lateral moraine, **210,** 212, 217
laterite. *See* tropical red soil
latitude, 234
lava, 7-9, 9, 14, 16-17, 19-20
leaching, 119-120, 164, 167
levée, 187-188
lightning, 50, **67**
lignite, 30
limestone, 24-27, 32, 93, **112**-113, 161, 171, 191
  landscapes, **191-194**
lip, of cirque, **208**
lode, of metal ore, **22**
loess, 116, 202
long profile, **178, 179, 180**
longitude, **234**
longshore drift, 227, 229
Los Angeles, California, **97,** 159
lowland
  glaciation, 214
  soils, 117
low pressure regions, 55-56, 61, 63-64, 68, 76

magma, 7, 9-11, 16, 18, 20-22, 44
mantle, 8
mass wasting, of soil, 158
meanders, **180-182,** 184
medial moraine, **210,** 212, 217
Mekong delta, **190**
meltwater, 212
meridian, **234**
mesquite bush, **125-126**
metal ores, 22
metamorphic rock, 10, 29, 44
meteorites, 6
meteorology, 50
mid-latitude "gears," **56**
migration of species, **134**
millibar, 63
minerals, 10, 22, 28
Mississippi delta, USA, 190
monoculture, 148

monsoon, 74, 76-79
moraine, 32, 116, 206-207, **210**, **212**-213, **215**
Moraine Lake, Alberta, **213**
mountain
   building, 40, 42-43, 45-46
   climates, **136-137**
   vegetation, **136-137**
Mount Baker, USA, 19
Mount Edziza, BC, 19
Mount Etna, Italy, 45
Mount Rainier, USA, **17**, 19
Mount St. Helens, USA, **5**, **11-13**, **16-17**, 19
mudflats, 189
mudflow, 13, **159**, 161, 167
mudpool, 21
mulch, 148
natural selection, 122-123
neap tide, 85
névé, 206, 213
Newfoundland, **43**, 46, **222**, **224**
Niagara escarpment, 27
Niagara Falls, **176**, **183**
Nile delta, **190**
Nile River, Egypt, 103-104, 188
nimbostratus, 58-59, 61
nomadic agriculture, **153**-154
normal fault, 33-34, 37
north
   geographic pole, 234
   magnetic pole, 234
northings, **234-235**
nutrients, 114-116, 120, 123, 131, 142, 145

oasis, **202**
occluded front, 68
ocean
   atmosphere, 94
   crust, **41**, 45
   currents, 54, **82-83**, 87
   floor, **41**, 45
   heating, 54, 81
   islands, 26-28, 42
   ridges, 44
   trenches, **41**-42, 44-45
   waves, 219
oil, 24, 29, 31-32, 46
Okanagan Lake, BC, **185**
Okanagan Valley, 97, **185**, 218
ores, **22**
outwash plains, 32, 216-217
overfold, 37
oxbow lake, 181

Pacific Ring of Fire, **10**-11, 40
pack ice, 83-**84**
palm oil plantation, **151**
Pangaea, **45-46**
peat, 117, 119
pebbles, 114
Pennsylvania, **2**-3, 36
permafrost, 167-**168**
physical weathering, 108, 110-111, 113-114, 116
plane routes, **70**

plankton, 87
plantation agriculture, 149-**151**, 152
plate
   boundaries, 40
   collisions, **43**-44
   crustal, 40-41, 45
playa lake, 198-199
plunge pool, **177**
Po delta, **190**
plunging waves, 225-**226**
podzolization, 119
polar easterlies, **56**
pollution
   of atmosphere, 113
   of rivers, 100, 104
ponderosa pine, **126-127**
potash, **31**, 46
potholes, **178**
Prairies, **24**-25, 30-31, 80, 123, 130-131, 142-143, 215
precipitation, 58-59
pressure, of air, 55
Prince George, climate graph, **73**
Prince Rupert, climate graph, **73**
pumice, 17, 18
pyramidal peak, 208, **210**, **212-213**
pyroclastic material, 17-18

quarrying, **207**, 213
quartz, 10, 29
Quinsam Coal, **32**
Quintette Coal, **39**

radiation
   fog, 65, 68
   solar, 51, 53-54
rain, 58, 61
   drops, 58
   forest, 130
   gauge, **236**
   shadow, 60
   splash, **163**, 167
range of species, 124
rapids, 177
Red River, Manitoba, **89-90**, 92, 94
Red Sea, 44-45
redwood (sequoia) tree, 126
reg, 201
relief
   cloud, 60
   precipitation, 60
   soils, 120
reservoir, 97, **99-100**, 185
reverse fault, 37
Rhine
   rift valley, Germany, **37**, **53**, **181**
   delta, **188**
ria, 230, 232
ribbon lakes, 185, **210**, **212**-214
ridge, oceanic, 44
rift valley, 34, 37
rill, 145
Ring of Fire, **10**-11, 40
Rio de la Plata delta, **230**
river
   channel, 90
   control, 90, 103, 174

   flow, 89-90, 94, 172-174
   levées (dikes), 90, **187-188**
   pollution, **100**-101
   terraces, **184**, 186
roche moutonnée, **208**, 213
rock basin lakes. *See* ribbon lakes
rock fall, 167, **169**
Rocky Mountain Trench, **38**, 46
Rocky Mountains, 24, 27, 30, 35, 46
rotation
   of earth, 54
   of crops, 142
rubber tree plantation, **152**

Sahara Desert, Africa, 199
Sahel, Africa, 145, 153
salt
   lakes, 198
   marsh, 228
   rock (halite), 25, 31
   water, 81
San Andreas Fault, USA, **34**, 37
San Francisco, 34
sand
   dunes, **199**, 228
   sea, **199**
   spit, **228**, **232**
sandstone, 10, 25, 93
Saskatchewan, 30-**31**, 80
Saskatoon water budget, **98**
saturation, 57, 60
savanna, 130-131, 154
scarp, 2, **171**
Scotland, **34-35**, 43
scree, 109, 114, **119**, **158**, 167, **169**, 217
sea breeze, 55
sea cliff tunnel, 225
seasons, **72**
sediment, 10, 23, **25**, 27, 29, 41, 44, 46, 86, **100**, 123, 173
sedimentary rock, 10, **23-25**, 29, 35, 112, 160, 171
seepage, 92
shale, 10, 25, **112**
shatter belt, South Australia, **38**
Ship Rock, New Mexico, USA, **15**
Sierra Nevadas, USA, **15**, 112
sill, 9-11, **15-16**
silt, 114
slash and burn agriculture, 129, **140-142**
sleet, 58
slumps, 167
smog, 65
snow, 57-58, 205-**206**
soil, 107, 114-120
   animals, 115
   components, 114
   conservation, 146, **147**, 148
   creep, **162**, 166-167
   erosion, 143-149
   exhaustion, 145
   fertility, 142
   formation, 114, 116, 120
   groups. *See* individual soil names

humus, 114-118
  layers, 116
  lowland, 117
  moisture, 91
  nutrients, 114-116, 120, 131, 142, 145
  organisms, 114, 116
  parent materials, 116
  pores, 115
  solution, 119, 164, 167
  upland, 118
  water, 114
solar energy, 51
  radiation, 53
solstice, 72, 76
solution, 164, 167, 175, 191-192
south
  geographic pole, 234
  magnetic pole, 234
spilling waves, **226**
spinning, earth, 56
spit. *See* sand spit
springs, 92, 93, **171**
  hot, 21
  limestone, **171**
  tides, 85
stack, 221-222, 225
stalactites, 192-**193**
stalagmites, 192-**193**
steppe, 131, 145
Stevenson's screen, 235-**236**
stone pavements, **201**
storm waves, 219
stratus, 58-59, 61
stream flow, 89-90, 94
striations, 207-208, 213
subduction, 46
Sudan, Africa, **1, 104, 154**
Suez, Egypt, **196**
Sullivan Mine, BC, **22**
sunshine recorder, **237**
surf, 227
surface wash, 163, 167
survival of species, 127
suspension, 173, 175
swallow hole, 192-**193**
swash, 223, **227**
swell wave, 219, 226
syncline, 36-37

tarn, 211
tear fault, 34, 37
Tennessee River, 98
terminal moraine, 212, **215-216**, 217
terraces, 184, **186**-187, 218
terracettes, 160, 167
thermometer, 235, **236**
thunder, 50, **66-67**
tidal **bore**, 85
tides, 85, 101
Tigris-Eurphrates delta, **190**
till, 116, 212-215, 217
Titanic, 84, 86, 88
tombolo, 232
Tonga-Kermadec trench, 41
tornado, 54
trade winds, 56
transpiration, 92, 124
trellis pattern, **171**
tropical
  rain forest, 127, 130, 142
  red soil, **118**, 120
tropics, 54, 58, 72, 76
truncated spur, 210-213
tundra
  climate, 79
  vegetation, 132
turbulence
  in atmosphere, 53
  in oceans, 81, **86-87**
typhoon. *See* hurricane

undercut cliff, 221, 225
undercutting, 221
upland
  glaciation, 203, **208-213**
  soils, 118, **119**
urban water movement, 95
U-shaped valley, **210**-211, 213

valley
  glacier, **210**, 212 213
  river, 188, **210**
  shape, **188**
Vancouver, BC (water budget), **98**
vegetation, 107, 121, **129**, 197
vein, metal ore, 22
Vesuvius, Italy, 14, 45
Victoria Falls, Africa, 176

viticulture (Rhine Valley), 52-**53**
volcano
  benefits of, 20
  disasters, **11-13, 19**
  distribution, **10**-11, 19, 34, **40-41**, 44-**45**
  dormant, 14
  eruptions, **5**, 7, 8, 11, 14, 16, 18, 41
  materials from, **16, 18**, 29, 113
  plug, **15**
wadi, 196, 198
warm air sector, 68
  front, 68
  ocean currents, 82-83
waste disposal, 168
water
  artesian, **93**
  conservation of, 90, **99**
  cycle, 95-96, 101, 170
  demand, 96-97
  droplets, 58
  power, **172**, 183
  sub-artesian, 150
  vapour, 51, 57
waterfalls, 175, **176-177**, 178, 183, **210**
wave-cut platform, 221, 223, **225**, 231-**232**
waves, 81, **219**-220, 226
weather, 50, 71
  charts, 63, **68**
  map symbols, **64**
  systems, 63
weathering, 108, 113, 157, 175
weir (dam), 237
Wensleydale, UK, **194**
westerly winds, 56
wet-bulb thermometer, 235
Whistler, BC, 205
wind, 56
wind erosion, accelerated, 144-145, 147
Winnipeg, Manitoba, **89-90**, 94

xerophytic plants, **125, 132-133**

Yellowstone, USA, 21, 27
Yosemite Valley, USA, **15, 37-38**, 175

KEYS FOR MULTIPLE-CHOICE QUESTIONS

PART 1: page 48: 1C; 2B; 3A; 4B; 5D; 6A; 7B; 8D; 9C; 10C; 11D

PART 2: page 105: 1B; 2C; 3A; 4C; 5B; 6A; 7B; 8C.

PART 3: page 156: 1C; 2D; 3B; 4A; 5B; 6A; 7D; 8C; 9D; 10A.

PART 4: page 233: 1A; 2D; 3A; 4A; 5D; 6D; 7B; 8A; 9D; 10B.

Additional sources and credits

4.9 From illustration by Bunji Tagawa from "The Changing Level of the Sea," by Rhodes W. Fairbridge. Copyright © 1960 by Scientific American Inc. All rights reserved. 15.6 Reproduced from the 1982 Ordnance Survey 1: 50 000 Landranger Map No. 98 with the permission of the Controller of Her Majesty's Stationery Office © Crown Copyright.

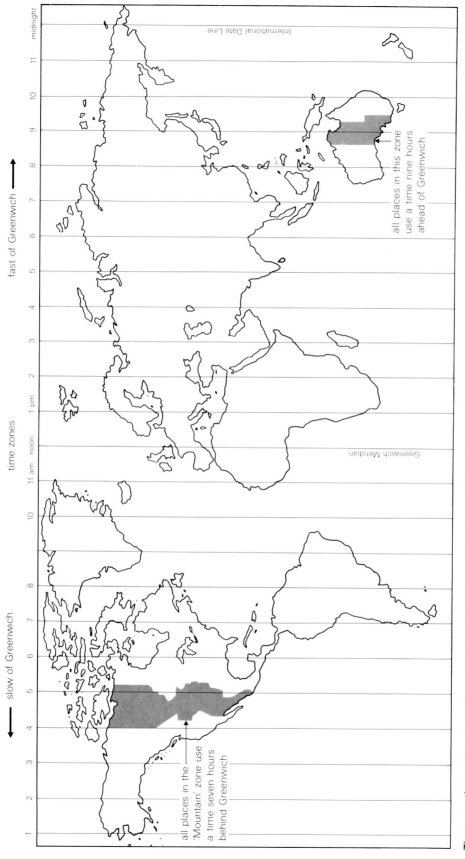

These are time zones around the world related to Universal Co-ordinated Time (Greenwich mean time). The time zones used by many countries do not correspond exactly with this scheme. Some examples are shown. Can you suggest practical reasons for these broad time bands?